An Introduction to the Theory of Distributions

PURE AND APPLIED MATHEMATICS

A Series of Monographs and Textbooks

1. K. Yano. Integral Formulas in Riemannian Geometry (1970)
2. S. Kobayashi. Hyperbolic Manifolds and Holomorphic Mappings (1970)
3. V. S. Vladimirov. Equations of Mathematical Physics (A. Jeffrey, editor; A. Littlewood, translator) (1970)
4. B. N. Pshenichnyi. Necessary Conditions for an Extremum (L. Neustadt, translation editor; K. Makowski, translator) (1971)
5. L. Narici, E. Beckenstein, and G. Bachman. Functional Analysis and Valuation Theory (1971)
6. D. S. Passman. Infinite Group Rings (1971)
7. L. Dornhoff. Group Representation Theory (in two parts). Part A: Ordinary Representation Theory. Part B: Modular Representation Theory (1971, 1972)
8. W. Boothby and G. L. Weiss (eds.). Symmetric Spaces: Short Courses Presented at Washington University (1972)
9. Y. Matsushima. Differentiable Manifolds (E. T. Kobayashi, translator) (1972)
10. L. E. Ward, Jr. Topology: An Outline for a First Course (1972)
11. A. Babakhanian. Cohomological Methods in Group Theory (1972)
12. R. Gilmer. Multiplicative Ideal Theory (1972)
13. J. Yeh. Stochastic Processes and the Wiener Integral (1973)
14. J. Barros-Neto. Introduction to the Theory of Distributions (1973)
15. R. Larsen. Functional Analysis: An Introduction (1973)
16. K. Yano and S. Ishihara. Tangent and Cotangent Bundles: Differential Geometry (1973)
17. C. Procesi. Rings with Polynomial Identities (1973)
18. R. Hermann. Geometry, Physics, and Systems (1973)
19. N. R. Wallach. Harmonic Analysis on Homogeneous Spaces (1973)
20. J. Dieudonné. Introduction to the Theory of Formal Groups (1973)
21. I. Vaisman. Cohomology and Differential Forms (1973)
22. B.-Y. Chen. Geometry of Submanifolds (1973)
23. M. Marcus. Finite Dimensional Multilinear Algebra (in two parts) (1973)
24. R. Larsen. Banach Algebras: An Introduction (1973)

In Preparation:

K. B. Stolarsky. Algebraic Numbers and Diophantine Approximation

AN INTRODUCTION TO THE THEORY OF DISTRIBUTIONS

José Barros-Neto

Department of Mathematics
Rutgers, The State University
New Brunswick, New Jersey

MARCEL DEKKER, INC. New York 1973

MARCEL DEKKER, INC.

95 Madison Avenue, New York, New York 10016

LIBRARY OF CONGRESS CATALOG CARD NUMBER: 72-90371

ISBN: 0-8247-6062-X

PRINTED IN THE UNITED STATES OF AMERICA

To

IVA,

CARMEN,

CLAUDIA,

MARILIA,

and

ANDRE

PREFACE

The present text corresponds to courses we taught at the universities of Montreal, Sao Paulo, and Rutgers.

It contains the spaces of distributions, distributions with compact support and tempered distributions; also their main properties and theorems of structure. Convolutions of distributions, convolutions of functions and distributions, regularization of distributions and convolutions maps are discussed in Chapter 3. Chapter 4 is devoted to the study of Fourier transforms of tempered distributions and their main properties, including the very important Paley-Wiener-Schwartz theorem. The aim of Chapter 6 is twofold: First, to prove the theorem of structure of $\mathcal{S}'$, the space of tempered distributions, and second to define the space $\mathcal{O}'_c$ of distributions rapidly decreasing at infinity, which is the space of convolution with $\mathcal{S}'$. As applications of the theory we expose, in Chapter 5, the L^2 theory of Sobolev spaces. Chapter 7 contains, among other things, a characterization of hypoelliptic partial differential operators with constant coefficients in terms of properties of their fundamental solutions, as well as Malgrange's original proof on the existence of fundamental solutions of partial differential operators with constant coefficients.

There are several intuitive or elementary ways of defining distributions—but they all lack the power, elegance, and flexibility of L. Schwartz's original presentation as a duality theory of topological vector spaces. Once the general theory of topological vector spaces is at our disposal, we can derive, as its most important application, the theory of distributions with all its properties and far-reaching consequences.

Since this is an introductory book, we content ourselves to develop, in Chapter 1, only those results on the theory of locally convex topological vector spaces that are needed in the subsequent chapters.

A student who has taken courses on advanced calculus, linear algebra, general topology (mostly metric spaces), and knows some of Lebesgue integration theory and Banach and Hilbert spaces (the Hahn-Banach theorem included), will have no difficulties in reading this book. The problems proposed at the end of every chapter serve him to check whether he has understood the subject matter.

We hope that, besides initiating the student to a very important theory, this book will motivate him to further his studies on the theories of topological vector spaces, distributions, and kernels, as well as their most important applications to analysis.

Princeton, N. J. J. B. N.

CONTENTS

An Introduction to the Theory of Distributions

Chapter 1

LOCALLY CONVEX SPACES

1. NOTATIONS AND TERMINOLOGY

Let $\underline{R}^n$ be the n-dimensional Euclidean space and let $x = (x_1, \cdots, x_n)$ be a variable element of $\underline{R}^n$. Its Euclidean norm will be denoted by

$$|x| = \left(\sum_{j=1}^{n} x_j^2\right)^{\frac{1}{2}}$$

Let $\underline{N}^n$ be the set of all n-tuples $p = (p_1, \cdots, p_n)$ with $p_j \in \underline{N}$, $1 \leq j \leq n$, where $\underline{N}$ denotes the set of all nonnegative integers. For all $p \in \underline{N}^n$, we set $|p| = p_1 + \cdots + p_n$. If $p, q \in \underline{N}^n$, define $p + q = (p_1 + q_1, \cdots, p_n + q_n)$. The notation $p \leq q$ means that $p_j \leq q_j$, $1 \leq j \leq n$. We set

$$p! = p_1! \cdots p_n!$$

as well as

$$\binom{p}{q} = \frac{p!}{q!\,(p-q)!} = \binom{p_1}{q_1}\binom{p_2}{q_2}\cdots\binom{p_n}{q_n}$$

If $x \in \underline{R}^n$ and $p \in \underline{N}^n$ then x^p is a notation for $x_1^{p_1} \cdots x_n^{p_n}$. According to these notations a polynomial in n variables $(x_1, \cdots, x_n)$, of degree $\leq m$ can be written as

$$p(x) = \sum_{|p| \leq m} a_p x^p$$

The coefficients a_p can be either complex numbers, in which case P is said to be a *polynomial with constant coefficients*, or complex-valued functions, in which case P will be a *polynomial with variable coefficients*.

Partial derivatives will be denoted by $\partial_j = \partial/\partial x_j$, $1 \leq j \leq n$, and we write

$$\partial = \frac{\partial}{\partial x} = \left(\frac{\partial}{\partial x_1}, \cdots, \frac{\partial}{\partial x_n}\right) .$$

We also set $D_j = (1/i)\partial/\partial x_j$, $1 \leq j \leq n$, where $i = \sqrt{-1}$, and we write

$$D = (D_1, \cdots, D_n).$$

If $p \in \underline{N}^n$, the notations

$$\partial^p \text{ or } \left(\frac{\partial}{\partial x}\right)^p$$

represent the partial derivative

$$\frac{\partial^{p_1 + \cdots + p_n}}{\partial x_1^{p_1} \cdots \partial x_n^{p_n}} .$$

Similarly, we set

$$D^p = D_1^{p_1} \cdots D_n^{p_n}.$$

According to these notations, a partial differential operator of order $\leq m$ can be written as follows:

$$P\left(x, \frac{\partial}{\partial x}\right) = \sum_{|p| \leq m} a_p(x) \left(\frac{\partial}{\partial x}\right)^p \text{ or } P(x, \partial) = \sum_{|p| \leq m} a_p(x) \partial^p,$$

where $a_p(x)$ are complex-valued functions defined on some open set of $\underline{R}^n$. When the coefficients a_p are constants we simply write

$$P\left(\frac{\partial}{\partial x}\right) = \sum_{|p|\le m} a_p \left(\frac{\partial}{\partial x}\right)^p \quad \text{or} \quad P(\partial) = \sum_{|p|\le m} a_p \partial^p.$$

Later on, when dealing with Fourier transforms, we shall consider partial differential operators

$$P(D) = \sum_{|p|\le m} a_p D^p$$

in the variables $D = (D_1,\cdots, D_n)$ rather than $\partial = (\partial_1,\cdots, \partial_n)$.

Definition 1.1. *Let* Ω *be an open set in* $\underline{R}^n$. *We denote by* $C^m(\Omega)$, *where* m *is a nonnegative integer, the vector space over* $\underline{C}$ *of all complex-valued functions defined in* Ω *having continuous derivatives of order* $\le$m. *We denote by* $C^\infty(\Omega)$ *the space of all complex-valued functions defined on* Ω *having derivatives of all orders.*

Clearly,

$$C^\infty(\Omega) = \bigcap_{m\ge 0} C^m(\Omega).$$

The elements of $C^\infty(\Omega)$ are called *infinitely differentiable functions* or C^∞ *functions* on Ω.

The General Leibniz Formula

Let $P = P(\partial)$ be a partial differential operator and let u and v be two functions belonging, say, to $C^\infty(\Omega)$. We have the following very useful formula:

$$P(\partial)(uv) = \sum_\alpha \frac{1}{\alpha!} \partial^\alpha u \cdot P^{(\alpha)}(\partial)v \qquad (1.1)$$

where $P^{(\alpha)}(\partial)$ is the partial differential operator obtained from the polynomial

$$P^{(\alpha)}(\eta) = \left(\frac{\partial}{\partial\eta}\right)^{\alpha} P(\eta)$$

by replacing the variable η by ∂. Indeed, by Taylor's formula, we have

$$P(\xi + \eta) = \sum_{\alpha} \frac{1}{\alpha!} \xi^{\alpha} P^{(\alpha)}(\eta). \tag{1.2}$$

On the other hand, by using the Leibniz rule

$$\partial_j(u\cdot v) = \partial_j u\cdot v + u\cdot\partial_j v,$$

we get

$$P(u\cdot v) = \sum_{\alpha} \partial^{\alpha} u\cdot R_{\alpha}(\partial)v \tag{1.3}$$

where $R_{\alpha}(\partial)$ is a partial differential operator. By setting $u = e^{<x,\xi>}$ and $v = e^{<x,\eta>}$ in (1.3), where $<x, \xi> = x_1\xi_1 +\cdots+ x_n\xi_n$ and $<x, \eta> = x_1\eta_1 +\cdots+ x_n\eta_n$, we get

$$P(\xi + \eta) = \sum_{\alpha} \xi^{\alpha}\cdot R_{\alpha}(\eta) \tag{1.4}$$

By comparing (1.2) and (1.4), we obtain

$$R_{\alpha}(\eta) = \frac{P^{(\alpha)}(\eta)}{\alpha!}$$

which proves (1.1).

2. TEST FUNCTIONS; REGULARIZATION

Definition 1.2. *Let* f *be a complex-valued function defined on an open subset* $\Omega \subset \underline{R}^n$. *We call support of* f, *and denote it by* supp f, *the closure in* Ω *of the set*

$$\{x \in \Omega: f(x) \neq 0\}.$$

The support of f is then the smallest relative closed set outside of which f is identically zero.

A *test function* on Ω is, by definition, a C^∞ function on Ω having *compact* support in Ω. The vector space of all test functions on Ω will be denoted by $C_c^\infty(\Omega)$ (or $\mathcal{D}(\Omega)$ in Schwartz's notation [28]). As we shall see in Chapter 2, this vector space equipped with an appropriate topology plays an important role in the definition of distributions on Ω. The following function

$$\beta(x) = \begin{cases} [\exp(|x|^2 - 1)]^{-1} & \text{if } |x| < 1 \\ 0 & \text{if } |x| \geq 1 \end{cases}$$

is a C^∞ function having for support the closed unit ball. By dividing by a constant, namely the integral of β over $\underline{R}^n$, we obtain another C^∞ function, denoted by α, with support the closed unit ball and such that

$$\int_{\underline{R}^n} \alpha(x)\, dx = 1.$$

Next, for every $\varepsilon > 0$, define

$$\alpha_\varepsilon(x) = \varepsilon^{-n}\alpha\left(\frac{x}{\varepsilon}\right).$$

Clearly $\alpha_\varepsilon(x) \in C_c^\infty(\underline{R}^n)$, its support is $\overline{B_\varepsilon(0)}$, the closed ball with center at the origin and radius ε and

$$\int_{\underline{R}^n} \alpha_\varepsilon(x)\, dx = 1.$$

With the help of the family of test functions $(\alpha_\varepsilon)_{\varepsilon>0}$, we can *regularize* discontinuous functions, like integrable ones, L^p functions, etc., that is to say, we can prove that such functions can be approximated by test functions. This is the aim of Theorem 1.1 below.

Definition 1.3. *A function* f *defined in* Ω *is said to be locally integrable if* f *is integrable (in the sense of Lebesgue) on every compact subset* $K \subset \Omega$

Equivalently, f is locally integrable in Ω if, for every compact set $K \subset \Omega$, the product $f \cdot \chi_K$ is integrable on Ω, where χ_K is the *characteristic function* of K, equal to 1 on K and 0 outside of K.

Definition 1.4. *Let* u *be a locally integrable function on* $\underline{R}^n$. *The function*

$$u_\varepsilon(x) = \int_{\underline{R}^n} u(x - y)\alpha_\varepsilon(y)\, dy = \int_{\underline{R}^n} u(y)\alpha_\varepsilon(x - y)\, dy \quad (1.5)$$

is said to be the convolution of u *and* α_ε. *It is also denoted by* $(u*\alpha_\varepsilon)(x)$ *or* $(\alpha_\varepsilon*u)(x)$.

Theorem 1.1 *Let* u *be a locally integrable function in* $\underline{R}^n$. *We have:*

1. *The convolution* u_ε *is a* C^∞ *function in* $\underline{R}^n$.

2. *If* u *has compact support* K *(and in the case* u *is integrable in* $\underline{R}^n$*), the support of* u_ε *is contained in the* ε*-neighborhood of* K.

3. *If* u *is a continuous function,* $u_\varepsilon \to u$ *uniformly on compact subsets of* $\underline{R}^n$.

4. *If* $u \in L^p(\underline{R}^n)$, $1 \leq p < +\infty$, *then* $u_\varepsilon \to u$ *in* $L^p(\underline{R}^n)$.

Proof. 1. In order to prove that $u_\varepsilon \in C^\infty(\underline{R}^n)$, it suffices to observe that both integrals in (1.5) are geing evaluated over compact subsets of $\underline{R}^n$ and to apply the classical theorem about differentiation inside the integral sign.

2. The ε-neighborhood of K is, by definition, the set

$$K_\varepsilon = \bigcup_{x \in K} B_\varepsilon(x)$$

of all closed balls with center $x \in K$ and radius ε. If $x \in \underline{R}^n$ and $d(x, K) > \varepsilon$ then $x \notin K_\varepsilon$ so that $\alpha_\varepsilon(x - y) = 0$ for all $y \in K$. Therefore, the second integral in (1.5) is equal to zero, which implies that the support of u_ε is contained in K_ε.

3. Suppose that u is a continuous function and let L be an arbitrary fixed compact subset of $\underline{R}^n$. Write

$$u_\varepsilon(x) - u(x) = \int_{\underline{R}^n} [u(x - y) - u(x)]\alpha_\varepsilon(y)\, dy.$$

Since u is *uniformly continuous* on L, given $\sigma > 0$, there is $\delta > 0$ such that

$$|u(x - y) - u(x)| < \sigma$$

for all $x \in L$ and for all $|y| < \delta$. By taking $\varepsilon \le \delta$, we get

$$|u_\varepsilon(x) - u(x)| \le \int_{\underline{R}^n} |u(x - y) - u(x)|\alpha_\varepsilon(y)\, dy < \sigma,$$

for all $x \in L$, which proves that $u_\varepsilon \to u$, uniformly on L, as $\varepsilon \to 0$.

4. Finally, suppose that $u \in L^p(\underline{R}^n)$, $1 \le p < +\infty$. It is known u can be approximated *in* L^p, by continuous functions with compact support [27, p. 68]. On the other hand, if $u_\varepsilon \in L^p$, then $u_\varepsilon \in L^p$. This follows from Minkowski's inequality in its integral form [33]:

$$\|u_\varepsilon\|_p = \left(\int_{\underline{R}^n} |u_\varepsilon(x)|^p \, dx\right)^{1/p} = \left(\int_{\underline{R}^n} \left|\int_{\underline{R}^n} u(x - y)\alpha_\varepsilon(y) \, dy\right|^p dx\right)^{1/p}$$

$$\le \int_{\underline{R}^n} \left\{\int_{\underline{R}^n} |u(x - y)\alpha_\varepsilon(y)|^p \, dx\right\}^{1/p} dy$$

$$= \int_{\underline{R}^n} \alpha_\varepsilon(y)\left\{\int_{\underline{R}^n} |u(x - y)|^p \, dx\right\}^{1/p} dy = \|u\|_p. \tag{1.6}$$

Given $\sigma > 0$, let v be a continuous function with compact support, such that

$$\|u - v\|_p < \frac{\sigma}{3}. \tag{1.7}$$

From inequalities (1.6) and (1.7) we get

$$\|u_\varepsilon - v_\varepsilon\|_p \le \|u - v\|_p < \frac{\sigma}{3}.$$

Next, write

$$\|u_\varepsilon - u\|_p \le \|u_\varepsilon - v_\varepsilon\|_p + \|v_\varepsilon - v\|_p + \|v - u\|_p. \tag{1.8}$$

Since v is a continuous function with compact support, then by part 3, $v_\varepsilon \to v$ uniformly on $\underline{R}^n$, hence $v_\varepsilon \to v$ in L^p. If we choose ε so small that

$$\|v_\varepsilon - v\|_p < \frac{\sigma}{3}$$

Then every one of the three terms at the right-hand side of the inequality (1.8) is smaller than $\sigma/3$, therefore $\|u_\varepsilon - u\|_p < \sigma$. Q.E.D.

Theorem 1.1 motivates the following definition.

Definition 1.5. *The family of test functions* $(\alpha_\varepsilon)_{\varepsilon>0}$ *is said to be a regularizing family of functions in* $\underline{R}^n$.

If, in particular, $\varepsilon = (j)^{-1}$, the sequence of functions

$$\alpha_j(x) = j^n\alpha(jx),\ j = 1, 2, \cdots,$$

is called a *regularizing sequence of functions*.

Corollary 1. *Let* Ω *be an open set in* $\underline{R}^n$. *Then* $C_c^\infty(\Omega)$ *is a dense subspace of* $L^p(\Omega)$, $1 \le p < +\infty$.

Corollary 2. *Let* K *be a compact set contained in* Ω. *There is a function* $\phi \in C_c^\infty(\Omega)$ *such that* $0 \le \phi \le 1$ *and* $\phi = 1$ *on* K.

Proof. Without loss of generality, we may assume Ω to be bounded. Let δ be the distance from K to the boundary of Ω and let $K_{\delta/3}$ be the $\delta/3$-neighborhood of K. It is easy to see that the function

$$\phi = \chi_{\delta/3} * \alpha_{\delta/3},$$

where $\chi_{\delta/3}$ denotes the characteristic function of $K_{\delta/3}$, satisfies the required properties. Q.E.D.

With a similar proof one can show that, if K is a compact subset in $\underline{R}^n$ and V is an arbitrary neighborhood of K, there is a function $\Phi \in C_c^\infty(\underline{R}^n)$ such that, $0 \le \phi \le 1$, ϕ is equal to 1 on a neighborhood of K and supp $\phi \subset V$.

Corollary 3. *Let* A *and* a *be real numbers such that* $0 < a < A$ *and denote by* B_A *and* B_{A-a} *concentric balls of radius* A *and* A-a, *respectively. There is a function* $\phi \in C_c^\infty(\underline{R}^n)$ *such that:* (i) supp $\phi \subset B_A$, (ii) $\phi(x) = 1$ *on* B_{A-a}, (iii) *for all* $p \in \underline{N}^n$, $|\partial^p\phi(x)| \le C(p, n)\cdot a^{-|p|}$, $\forall\, x \in \underline{R}^n$.

Proof. Let χ be the characteristic function of the concentric ball with radius A - (2a/3) and define

$$\phi(x) = \chi * \alpha_\delta(x) = \int_{B_{A-(2a/3)}} \alpha_\delta(x - y)\, dy = \frac{1}{\delta^n} \int_{B_{A-(2a/3)}} \alpha\left(\frac{x-y}{\delta}\right) dy,$$

with $\delta = a/3$. It is clear that supp $\phi \subset B_A$ and that $\phi = 1$ on B_{A-a}. For all $1 \le j \le n$, we have

$$\partial_j\phi(x) = \frac{\delta^{-1}}{\delta^n} \int_{B_{A-(2a/3)}} \frac{\partial\alpha}{\partial x_j} \left(\frac{x-y}{\delta}\right) dy,$$

hence

$$|\partial_j\phi(x)| \le \frac{\delta^{-1}}{\delta^n} \int_{\underline{R}^n} \partial_j\alpha \left(\frac{y}{\delta}\right) dy = \delta^{-1} \int_{\underline{R}^n} \partial_j\alpha(t)\, dt \le C(j, n)\cdot a^{-1}.$$

With a similar proof we get condition (iii). Q.E.D.

Corollary 4. *Let* K *be a compact subset of* $\underline{R}^n$ *and let* $(\Omega_j)_{1\le j\le k}$ *be a finite open covering of* K. *There are functions* $\phi_j \in C_c^\infty(\underline{R}^n)$ *such that* $0 \le \phi_j \le 1$, $1 \le j \le k$, *and* $\sum_{j=1}^k \phi_j = 1$ *on a neighborhood of* K.

Proof. We can find compact sets $(K_j)_{1\le j\le k}$ such that $K_j \subset \Omega_j$ and

$$K \subset \bigcup_{j=1}^{k} \mathring{K}_j,$$

where $\mathring{K}_j$ denotes the interior of K_j. For every j, let $\psi_j \in C_c^\infty(\Omega_j)$ be such that $0 \le \psi_j \le 1$ and $\psi_j = 1$ on K_j. Setting

$$\phi_1 = \psi_1 \text{ and } \phi_j = \psi_j(1 - \psi_1) \cdots (1 - \psi_{j-1}),\ j = 2,\cdots, k,$$

it is easy to check that the functions $(\phi_j)_{1\leq j\leq k}$ satisfy the required properties. Q.E.D.

The functions $(\phi_j)_{1\leq j\leq k}$ are said to be a C^∞ *partition of unity subordinated to the covering* $(\Omega_j)_{1\leq j\leq k}$ *of* K.

3. SEMINORMS; LOCALLY CONVEX SPACES

Let E be a vector space over a field κ which we always assume to be $\underline{R}$, the field of real numbers, or $\underline{C}$, the field of complex numbers. Denote by $\underline{R}_+$ the set of all nonnegative real numbers.

Definition 1.6. *A seminorm on* E *is a map* $p: E \to \underline{R}_+$ *satisfying the following axioms:*

(i) $p(x + y) \leq p(x) + p(y), \forall\, x, y \in E$;

(ii) $p(\lambda x) = |\lambda|\, p(x), \forall\, x \in E, \forall\, \lambda \in K$.

We say that p *is a norm if, in addition, it satisfies the following axiom:*

$$p(x) = 0 \textit{ if and only if } x = 0.$$

Examples. 1. Let p be a real number such that $1 \leq p < +\infty$. The map

$$x = (x_1, \cdots, x_n) \quad \underline{R}^n \to \|x\| = \left(\sum_{i=1}^{n} |x_i|^p\right)^{1/p} \in \underline{R}_+$$

defines a norm on $\underline{R}^n$. Also

$$\|x\|_\infty = \sup_{1\leq i\leq n} |x_i|$$

is a norm in $\underline{R}^n$.

2. Let $I = [a, b]$ and let $C(I)$ be the vector space of all complex-valued continuous functions defined on I. If $f \in C(I)$ then

$$\|f\|_1 = \int_a^b |f(t)|\, dt \text{ and } \|f\|_\infty = \sup_{t\in I} |f(t)|$$

define norms on C(I).

3. Denote by $L^p(I)$, where $1 < p < +\infty$, the space of functions f defined on I and such that $|f|^p$ is integrable on I in the Lebesgue sense. It can be shown that

$$\|f\|_p = \left(\int_a^b |f(t)|^p dt\right)^{1/p}$$

is a seminorm on $L^p(I)$.

4. Let X be a topological space and let C(X) be the vector space of all complex-valued continuous functions on X. Let K be a compact subset of X and define

$$p_K(f) = \sup_{t\in K} |f(t)|.$$

For each K, the map $p_K\colon\ C(X) \to \underline{R}_+$ is a seminorm.

Locally Convex Spaces

Let E be a vector space over κ. We say that a topology defined on E is *compatible* with the vector space structure of E if the maps

$$(x, y) \in E \times E \to x + y \in E$$

$$(\lambda, x) \in K \times E \to \lambda x \in E$$

are continuous. A vector space equipped with a compatible topology is said to be a *topological vector space* (TVS). The topology of a TVS can be described in terms of a *fundamental system* (or *basis*) *of neighborhoods of the origin.* In a topological space a collection of sets is said to be a fundamental system of neighborhoods of a point if every set of the collection contains the point, the intersection of any two sets of the collection contains a set belonging to the collection, and every neighborhood of the point contains a set of the collection.

Since all the spaces we shall deal with will be *locally convex spaces* and for these a fundamental system of neighborhoods of the origin can be described more easily by a family of seminorms, we shall not discuss here the general definition of a topological vector space or the general properties of neighborhoods of zero. The reader should consult Refs. [6, 12, 17, and 19].

Let p be a seminorm on E. An *open (resp. closed) ball with center* $x_0 \in E$ *and radius* $r > 0$ *is the set*

$$B(x_0, r) = \{x \in E: \ p(x - x_0) < r\}$$

[resp. $\bar{B}(x_0, r) = \{x \in E: p(x - x_0) \leq r\}$]. A neighborhood of x_0 is a set V containing a ball with center x_0. It is very easy to see that the set of all open (closed) balls with center x_0, an arbitrary element of E, defines a *fundamental system of neighborhoods of* x_0 of a topology on E which is *compatible with the vector space structure* of E.

Furthermore, if a is a fixed element of E and if $\lambda \neq 0$ is a fixed element of κ, the maps

$$x \to a + x \qquad (\textit{translation})$$

and

$$x \to \lambda x \qquad (\textit{homothety})$$

are *homeomorphisms* on E. It follows that, in order to know a fundamental system of neighborhoods of a point $a \in E$, it suffices to know a fundamental system of neighborhoods of the origin.

Let $(p_i)_{i \in I}$ be a family of seminorms defined on E. For every $x_0 \in E$, ε a positive real number, and F a finite part of I, define

$$V(x_0, \varepsilon, F) = \{x \in E: p_i(x - x_0) < \varepsilon, \ i \in F\}.$$

The set $V(x_0, \varepsilon, F)$ is clearly the intersection of the balls with center x_0 and radius ε, corresponding to the seminorms p_i with $i \in F$.

The family of all sets $V(x_0, \varepsilon, F)$ when ε runs through the set of all positive real numbers and F runs through all finite subsets of I defines a fundamental system of neighborhoods of x_0 of a topology on E compatible with the vector space structure of E. Equipped with this topology, E is said to be a *locally convex topological vector space* (LCS).

If E is a LCS and $(p_i)_{i \in I}$ a family of seminorms defining its topology, then it is easy to show that E is a *Hausdorff space* if and only if to any pair x, y of elements of E with $x \neq y$ it corresponds to a seminorm p_k such that $p_k(x) \neq p_k(y)$.

Convex and Balanced Sets

Definition 1.7. *A subset* A *of a vector space* E *over* K *is said to be convex if, given two points* $x, y \in A$, *the segment*

$$\alpha x + \beta y$$

where $\alpha, \beta \in \mathbb{R}_+$ *and* $\alpha + \beta = 1$ *is contained in* A.

Examples. 1. The whole space E is convex; the empty set is convex.

2. Balls are convex sets.

3. Segments are convex sets.

4. If $(A_i)_{i \in I}$ is a family of convex sets of E, then the intersection

$$A = \bigcap_{i \in I} A_i$$

is a convex set.

Definition 1.8. *Let* A *be a subset of* E. *The convex hull of*

A *is the smallest convex subset of* E *containing* A.

The convex hull of a given set A is always well defined. Indeed, it suffices to take the intersection of the family of all convex sets containing A and to observe that this family is nonempty since it contains the whole space E.

We denote by $\Gamma(A)$ the convex hull of A. It can also be described as follows: $\Gamma(A)$ is the set of all the elements $x \in E$ that can be represented as a finite sum

$$x = \sum_{i \in F} \alpha_i x_i,$$

where $x_i \in A$, $\alpha_i \in \underline{R}_+$, $\sum_{i \in F} \alpha_i = 1$, and F a finite set of indices depending on x.

Definition 1.9. *A subset* A *of* E *is said to be balanced if* $\lambda A \subset A$ *for all* $\lambda \in K$ *such that* $|\lambda| \leq 1$.

Examples. 1. Balls with center at the origin are balanced sets.

2. Let $E = \underline{R}$. The set [0, 1] is a convex one, but it is not balanced.

Suppose that A is a convex and balanced subset of E. Then it is easy to verify that for every pair of points $x, y \in A$ we have

$$\alpha x + \beta y \in A$$

for all $\alpha, \beta \in K$ such that $|\alpha| + |\beta| \leq 1$.

Also, it is easy to check that the intersection of a family of convex and balanced subsets of E is a convex and balanced subset of E.

Definition 1.10. *Let* A *be a subset of* E. *The balanced convex hull of* A *is the smallest balanced convex subset* $\Gamma_b(A)$ *of* E *contain-* A.

Given A, it is clear that its balanced convex hull is the intersection of all balanced convex sets containing A. Also, $\Gamma_b(A)$ can be described as follows: $x \in \Gamma_b(A)$ if and only if

$$x = \sum_{i \in F} \alpha_i x_i$$

with $x_i \in A$, $\sum_{i \in F} |\alpha_i| \leq 1$, and F a finite set of indices.

Absorbing Sets

Definition 1.11. *A subset* V *of a vector space* E *is said to be absorbing if given* $x \in E$, *there is a real number* $\lambda > 0$ *such that* $\lambda x \in V$.

Examples. 1. The set {-1, 0, 1} is an absorbing set in the real line $\underline{R}$. In the same manner, the set consisting of a circumference with center at the origin plus the origin is an absorbing set in $\underline{R}^2$.

2. Balls with center at the origin are absorbing sets. More generally, every neighborhood of the origin in a topological vector space is an absorbing set.

Let E be a vector space equipped with a seminorm p. The *unit ball*

$$U = \{x \in E: p(x) < 1\}$$

satisfies the following properties: (i) U is a balanced convex set; (ii) U is absorbing.

Conversely, let us show that if V is a balanced convex and absorbing subset of E then

$$q(x) = \inf\{\lambda \geq 0: x \in \lambda V\}$$

is a seminorm on E.

Indeed, since V is absorbing, q is well defined on E and $q\colon E \to \underline{R}_+$. If x, y $\in$ E, let $\lambda > 0$ and $\mu > 0$ be such that

$$x \in \lambda V \text{ and } y \in \mu V.$$

We have

$$x + y \in \lambda V + \mu V = (\lambda + \mu)\left[\frac{\lambda}{\lambda+\mu} V + \frac{\mu}{\lambda+\mu} V\right] \subset (\lambda + \mu)V$$

because V is convex. This implies that

$$q(x + y) \leq q(x) + q(y).$$

Finally, the assumption that V is a balanced set implies that

$$q(\lambda x) = |\lambda| q(x). \text{ Q.E.D.}$$

The following relation is easy to show:

$$\{x \in E\colon q(x) < 1\} \subset V \subset \{x \in E\colon q(x) \leq 1\}.$$

We can now prove that the following theorem, which gives a characterization of locally convex spaces and at the same time justifies the name given to such spaces.

Theorem 1.2. *Let* E *be a topological vector space. The following are equivalent conditions:*

(1) E *is locally convex;*

(2) *There is a funduamental system of convex neighborhoods of the origin.*

(3) *There is a fundamental system of absorbing balanced convex neighborhoods of the origin.*

Proof. (1) => (2). If the topology of E is defined by a family of seminorms $(p_i)_{i \in I}$, then the sets

$$V(F, \varepsilon) = \{x \in E\colon p_i(x) \leq \varepsilon,\ i \in F\}$$

where F is a finite part of I and $0 < \varepsilon < 1$, form a fundamental system of convex neighborhoods of the origin.

(3) => (1). The every V an absorbing balanced convex subset of E belonging to a fundamental system of neighborhoods of the origin we associate the seminorm p_V. It is easy to see that the family of seminorms obtained in this way defines the topology of E.

(2) => (3). It suffices to show that if V is a convex neighborhood of the origin then the set

$$U = \bigcap_{|\lambda|=1} \lambda V$$

is an absorbing balanced convex neighborhood of the origin. Since the map $(\mu, x) \to \mu x$ is continuous at the point (0, 0), there are $\varepsilon > 0$ and V', a neighborhood of zero in E, such that

$$\mu x \in V \text{ for all } |\mu| \leq \varepsilon \text{ and for all } x \in V'.$$

Or, equivalently, there is a neighborhood of zero W such that

$$\mu W \subset V \text{ for all } |\mu| \leq 1.$$

The last relation implies that

$$\mu W \subset V \text{ for all } |\mu| = 1.$$

Hence $W \subset \lambda V$ for all $|\lambda| = 1$, which implies that U is a neighborhood of zero in E.

Clearly, U is a convex set (as an intersection of convex sets) and it is absorbing. Let us prove that U is balanced. If $x \in U$, the segment [0, x] is contained in U, i.e.,

$$\lambda x \in U \text{ for all } 0 \leq \lambda \leq 1.$$

On the other hand, if $x \in U$, it follows from the definition of U that $\lambda x \in U$ for all $|\lambda| = 1$. Therefore, if $\mu \neq 0$ and $|\mu| \leq 1$ we get

$$\mu x = |\mu| \frac{\mu}{|\mu|} x \in U$$

which proves that U is balanced. Q.E.D.

4. EXAMPLES OF LOCALLY CONVEX SPACES

1. A seminormed space is a locally convex space. In particular, a normed space is a locally convex one.

2. Let X be a locally compact topological space, let C(X) be the space of all complex-valued continuous functions on X, and let K denote the collection of all compact subsets of X. The family of seminorms $(p_K)_{K\in K}$, where $p_K(f) = \sup_{t\in K} |f(t)|$, defines a Hausdorff locally convex topology on C(X). It can be seen that a sequence (f_j) of functions of C(X) converges to zero in this topology if and only if $f_j(x)$ converges to zero *uniformly on each compact* K of X. For this reason, this locally convex topology is called the *topology of uniform convergence on compact subsets of* X.

In particular, if $X = \Omega$, an open subset of $\underline{R}^n$, such topology can be defined by a sequence of seminorms. In fact, it suffices to take an increasing sequence $(K_j)_{j\in N}$ of compact subsets contained in Ω whose union is Ω and the corresponding sequence of seminorms $(p_{K_j})_{j\in N}$. In this case, the topology of $C(\Omega)$ is defined by a *countable* fundamental system of neighborhood of the origin. The Hausdorff locally convex space $C(\Omega)$ is then a *metrizable space* [4, 18]. Moreover, $C(\Omega)$ is a *complete* space since, by a classical theorem, the uniform limit on compact subsets of Ω of continuous functions is a continuous function. We often refer to the topology just defined on $C(\Omega)$ as the *natural topology* of $C(\Omega)$.

Definition 1.12. *A Hausdorff locally convex, metrizable, and complete space is said to be a Frechet space.*

3. Let Ω be an open subset of $\underline{R}^n$ and let $1 \leq p < +\infty$. Denote by $L^p_{\ell oc}(\Omega)$ the space of classes of measurable functions such that on every compact subset K of Ω we have

$$\int_K \left(|f(x)|^p \, dx\right)^{1/p} < +\infty .$$

The elements of $L^p_{\ell oc}(\Omega)$ are called pth *power locally integrable functions*. Define the seminorm

$$p_K(f) = \left(\int_K |f(x)|^p \, dx\right)^{1/p}.$$

The family of seminorms (p_K), where K runs over all compact subsets of Ω, defines a locally convex topology on $L^p_{\ell oc}(\Omega)$. Such topology can be defined by a countable family of seminorms and it can be shown that $L^p_{\ell oc}(\Omega)$ is a Frechet space. In an analogous way, one defines the Frechet space $L^\infty_{\ell oc}(\Omega)$.

5. DUALS

Given E a vector space over a field , its *algebraic dual* E^* is the vector space of all linear maps $x^* : E^* \to \kappa$. We denote by

$$x^*(x) \text{ or } \langle x, x^* \rangle$$

the value of x^* on $x \in E$.

When E is a topological vector space over a field κ (that we always suppose equal to $\underline{R}$ or $\underline{C}$), the *dual* E' of E is the subspace of E^* consisting of *all continuous linear maps* (or *functionals*) on E.

If for every $x^* \in E^*$ we set

$$p_{x^*}(x) = |\langle x, x^* \rangle|$$

then p_{x^*} defines a seminorm on E and the family $(p_{x^*})_{x^* \in E^*}$ defines a locally convex topology on E, denoted by $\sigma(E, E^*)$. In a similar manner we can define the topology $\sigma(E^*, E)$ on E^*. When E is a topological vector space, the topology $\sigma(E, E')$ defined on E by the family of seminorms $(p_{x'})_{x' \in E'}$ is said to be the *weak* topology of E. Obviously, the weak topology is coarser than the given topology of E and also coarser than $\sigma(E, E^*)$.

In a similar way we can define the *weak topology* $\sigma(E', E)$ on the dual E' of E. It immediately follows that a sequence (x_j') converges weakly to zero in E' if and only if, for every $x \in E$, the sequence $(x_j'(x))$ converges to zero in κ. Therefore, the weak topology on E' coincides with the *pointwise convergence topology*.

On the dual E' of a topological vector space E we can also define another very important locally convex topology, namely the *strong topology* of E'. In order to define it we need the following.

Definition 1.13. *Let* E *be a topological vector space. We say that a subset* A *of* E *is bounded if, given* V *a neighborhood of zero in* E, *there is a number* $\lambda > 0$ *such that* $\lambda A \subset V$.

When E is a locally convex topological vector space, by Theorem 1.2, every neighborhood of zero contains a balanced neighborhood of zero; thus Definition 1.13 is equivalent to the following one: A *is bounded in* E *if to every neighborhood of zero* V *there is a number* $\varepsilon > 0$ *such that* $\lambda A \subset V$ for all $|\lambda| \le \varepsilon$.

We remark that the two definitions are also equivalent in the general case since it can be shown that every topological vector space has a fundamental system of balanced neighborhoods of zero.

Examples. 1. Finite subsets of E are bounded sets.

2. In a seminormed space, balls are bounded sets.

3. Every relatively compact subset A of a locally convex space E is bounded. Indeed, given V a neighborhood of zero in E, there is an open neighborhood of zero W in E such that $W + W \subset V$ and $\mu W \subset W$ for all $|\mu| \le 1$. Since A is relatively compact, we can find a finite subset $(x_j)_{1 \le j \le p}$ of elements of A such that the open sets $(x_j + W)_{1 \le j \le p}$ cover A. Since the set $(x_j)_{1 \le j \le p}$ is bounded in E, we can find a real number $0 < \lambda < 1$ such that $\lambda x_j \subset W$, $1 \le j \le p$. We then have

$$\lambda A \subset \bigcup_{j=1}^{p} \lambda(x_j + W) \subset W + W \subset V,$$

hence A is bounded. Q.E.D.

Definition 1.14. *Let* B *be a subset of a topological vector space* E. *The polar set* B° *of* B *is the subset of* E' *defined as follows:*

$$B^{\circ} = \{x' \in E' : |<x, x'>| \leq 1, \forall\, x \in B\}.$$

We are going to show that if A is a bounded subset of E, its polar set A° is an *absorbing, balanced,* and *convex* subset of E'. Indeed, if x', y' $\in$ A° and α, β are nonnegative real numbers such that $\alpha + \beta = 1$, we have

$$|<x, \alpha x' + \beta y'>| \leq \alpha|<x, x'>| + \beta|<x, y'>| \leq 1,$$

hence A° is convex.

If $x' \in A^{\circ}$ and $\lambda \in \kappa$ is such that $|\lambda| \leq 1$, we have

$$|<x, \lambda x'>| = |\lambda|\ |<x, x'>| \leq 1,$$

hence $\lambda x' \in A^{\circ}$ and, consequently A° is balanced.

Finally, let $z' \in E'$ and consider the following neighborhood of zero in E:

$$V = \{x \in E: |<x, z'>| \leq 1\}.$$

Since A is a bounded subset of E, there is a $\lambda > 0$ such that $\lambda A \subset V$; hence

$$|<x, \lambda z'>| = |<\lambda x, z'>| \leq 1, \forall\, x \in A,$$

which proves that A° is an absorbing set in E'.

Therefore, by the results of Section 3 preceding Theorem 1.2, to every bounded set A of E there corresponds the following seminorm on E':

$$p_{A^\circ}(x') = \inf \{\lambda \geq 0: x' \in \lambda A^\circ\}.$$

If $\mathcal{B}$ denotes the family of all bounded subsets of E, the family of seminorms $(p_{A^\circ})_{A \in \mathcal{B}}$ defines a Hausdorff locally convex topology on E', called the *strong topology* of E'. It can be shown that a sequence (x_j') converges strongly to zero in E' if and only if the sequence $(x_j'(x))$ converges uniformly to zero on every bounded set of E. For this reason, the strong topology on E' is also called the *topology of uniform convergence on bounded sets of* E. We denote by E_b' the dual E' equipped with the strong topology.

As an example, we mention that if E is a normed space, its dual E' equipped with the norm

$$\|x'\|_{E'} = \sup_{\|x\|_E \leq 1} |\langle x, x'\rangle|$$

becomes a *Banach space*. Then, the strong topology on E' coincides with the one defined by the above norm.

Reflexive Spaces

If we denote by E'' the dual of the Banach space E', then on E'' we can define the norm

$$\|x''\|_{E''} = \sup_{\|x'\|_{E'} \leq 1} |\langle x', x''\rangle|$$

which turns E'' into a Banach space.

Every element $x \in E$ defines a continuous linear functional $\tilde{x}$ on E' as follows:

$$\tilde{x}(x') = \langle x, x'\rangle, \ \forall \ x' \in E'.$$

It can be shown that the map $x \to \tilde{x}$ is an isometry from E into E''. If this isometry is *onto* we say that E is a reflexive space.

More generally, let E be a Hausdorff topological vector space and let E_b' be its strong dual. The *bidual* E'' of E is, by definition, the dual of E_b'. As before, we can define the map

$x \in E \to \tilde{x} \in E''$. We say that the topological vector space E is *reflexive* if this map is an isomorphism from E onto E''_b.

Examples. 1. The n-dimensional Euclidean space $\underline{R}^n$ is reflexive. More generally, finite-dimensional Hausdorff topological vector spaces are reflexive.

2. Hilbert spaces are reflexive spaces.

3. The Banach spaces $L^p(\Omega)$, $1 < p < +\infty$, are reflexive. The spaces $L'(\Omega)$ and $L^\infty(\Omega)$ are not reflexive.

6. THE INDUCTIVE LIMIT TOPOLOGY

We are going to define the inductive limit topology in a particularly simple case which will suffice for the definition of distributions to be discussed in the next chapter. For a more general discussion on inductive limits, the reader should consult Refs. 6, 12, 17, 19, and 32.

Let $(E_i)_{i \in \underline{N}}$ be an increasing sequence of locally convex spaces such that the identity map $E_i \to E_{i+1}$ is continuous for every i. Let

$$E = \bigcup_{i=1}^{\infty} E_i$$

and define on E the *finest* locally convex topology for which the identity map $E_i \to E$ is continuous for every $i = 1, 2, \cdots$. This topology is called the *inductive limit topology* of E defined by the subspaces E_i and the space E equipped with such topology is said to be *the inductive limit of the spaces* $(E_i)_{i \in \underline{N}}$.

In order that a convex set V be a neighborhood of zero in the inductive limit topology, it is necessary and sufficient that every intersection $V \cap E_i$ be a neighborhood of zero in E_i for all $i = 1, 2, \cdots$. Also, we get a fundamental system of neighborhoods of the origin in E by taking all convex hulls

$$V = \Gamma\left(\bigcup_{i=1}^{\infty} V_i\right)$$

where every V_i runs over a fundamental system of convex neighborhoods of zero in E_i, $i = 1, 2, \cdots$.

Proposition 1.1. *Let* E *be the inductive limit of* $(E_i)_{i \in N}$ *and let* F *be any locally convex space. A linear map* $u: E \to \overline{F}$ *is continuous if and only if the restriction* u_i *of* u *is continuous from* E_i *into* F *for all* i.

Proof. If us is continuous, then every restriction u_i is also continuous because, by definition, the identity map $E_i \to E$ is continuous. Conversely, suppose that every u_i is a continuous map from E_i into F. Given U a convex neighborhood of zero in F, there is a convex neighborhood of zero V_i in E_i such that $u_i(V_i) \subset U$. Then,

$$V = \Gamma\left(\bigcup_{i=1}^{\infty} V_i\right)$$

is a neighborhood of zero in E and, clearly $u(V) \subset U$; hence u is continuous from E into F. Q.E.D.

Theorem 1.3. *Suppose that a vector space* E *is the union of an increasing sequence* $(E_i)_{i \in N}$ *of locally convex spaces such that:*

(1) *For every* i *the identity map* $E_i \to E_{i+1}$ *is continuous.*

(2) *The topology induced by* E_{i+1} *on* E_i *coincides with the topology of* $E_i, \forall\, i$.

(3) E_i *is a closed subspace of* $E_{i+1}, \forall\, i$.

Then (i) *the inductive limit topology of* E *induces on every* E_i *its original topology;* (ii) *a subset* A *is bounded in the inductive limit* E *if and only if there is an index* j *such that* A *is contained and it is bounded in* E_j.

The proof is based on the following lemma.

Lemma 1.1. *Let* F *be a locally convex space and let* G *be a closed subspace of* F. *Suppose that* V *is a balanced convex open neighborhood of zero in* G *and let* $x \in F$ *be such that* $x \notin G$. *Then, there is a balanced convex open neighborhood of zero* W *in* F *such that* $x \notin W$ *and* $W \cap G = V$.

Proof. Since G is closed in F, there is a balanced convex open neighborhood of zero V_0 in F such that

$$(x + V_0) \cap G = \phi \text{ and } V_0 \cap G \subset V.$$

Let W be the balanced convex hull of $V \cup V_0$. It is easy to see that W is open. Let us prove that $W \cap G = V$. Clearly, $W \cap G \supset V$. If $w \in W \cap G$ we can write (see Problem 21)

$$w = \alpha v + \beta v_0$$

with $v \in V$, $v_0 \in V$, and $|\alpha| + |\beta| \leq 1$. We can assume that $\beta \neq 0$, otherwise there is nothing to prove. But then the above relation implies that $v_0 \in V_0 \cap G \subset V$, thus $w \in V$. Finally, suppose by contradiction that $x \in W$. Then, $x = y + z$ with $y \in G$ and $z \in V_0$. Hence,

$$y = x - z \in (x + v_0) \cap G$$

which is impossible. Q.E.D.

Proof of Theorem 1.3. 1. In order to show that the topology induced by E on every E_i coincides with the given topology of E_i, it suffices to show that, given V_i a balanced convex neighborhood of zero in E_i, there is a neighborhood of zero V in E such that

$$V_i = V \cap E_i, \forall\ i.$$

By applying the lemma, it is easy to see that there is a sequence

$$(V_{i+k}),\ k = 0, 1, 2, \cdots,$$

of balanced convex neighborhoods of zero in E_{i+k} such that

$$V_{i+k-1} = V_{i+k} \cap E_{i+k-1}, \quad k = 1, 2, \cdots.$$

If we set

$$V = \bigcup_{k=0}^{\infty} V_{i+k},$$

it is easy to see that V is a neighborhood of zero in E and that $V_i = V \cap E_i$.

2. Let A be a bounded set in E. Suppose, by contradiction, that there is no index i such that A is contained in E_i. Then, we can find an increasing sequence of indices (i_n) and a sequence (x_n) of elements of E such that

$$x_n \in A \cap E_{i_n} \text{ and } x_n \notin E_{i_{n-1}}.$$

By the lemma, there is a sequence (V_n) of balanced convex open neighborhoods of zero in E_{i_n} such that

$$x_n \notin nV_n \text{ and } V_n \cap E_{i_{n-1}} = V_{n-1}.$$

Let $V = \bigcup_1^{\infty} V_n$. Then V is a neighborhood of zero in E such that

$$V \cap E_{i_n} = V_n \text{ and } x_n \notin nV.$$

But this contradicts our assumption that A is bounded E. Q.E.D.

Examples. 1. Let Ω be an open set in $\underline{R}^n$ and let (K_i) be an increasing sequence of compact subsets of Ω such that $\Omega = \bigcup_i K_i$. Let

$$E = C_c(\Omega)$$

be the space of all complex-valued continuous functions defined in Ω and with compact support. Let

$$E_i = C_c(\Omega;\ K_i)$$

be the subspace of $C_c(\Omega)$ consisting of all functions having support in K_i. It is clear that

$$C_c(\Omega) = \bigcup_i C_c(\Omega;\ K_i).$$

Define on $C_c(\Omega;\ K_i)$ the topology of uniform convergence on K_i (see Example 2 in Section 4). Such topology is a locally convex one defined by the seminorm

$$p_{K_i}(f) = \sup_{x \in K_i} |f(x)|.$$

It is easy to see that p_{K_i} is not only a seminorm but actually a norm on $C_c(\Omega;\ K_i)$ and that equipped with this norm $C_c(\Omega;\ K_i)$ is a Banach space. We leave to the reader the verification that the imbedding

$$C_c(\Omega;\ K_i) \to C_c(\Omega;\ K_{i+1})$$

is continuous. Also, $C_c(\Omega;\ K_i)$ is a closed subspace of $C_c(\Omega;\ K_{i+1})$. All the assumptions of Theorem 1.3 being verified, we define on $C_c(\Omega)$ the inductive limit topology of the spaces $C_c(\Omega;\ K_i)$. Such topology will be called the *natural topology* of $C_c(\Omega)$.

As a consequence of Theorem 1.3, we can see that *a sequence* (f_j) *converges to zero in* $C_c(\Omega)$ *if and only if the following conditions are satisfied:* (i) *there is a compact subset* K *of* Ω *such that* $\operatorname{supp} f_j \subset K$, *for every* j; (ii) *the sequence* (f_j) *converges to zero uniformly on* K.

Definition 1.15. *A Radon measure on* Ω *is a continuous linear functional on* $C_c(\Omega)$.

The space of all Radon measures on Ω will be denoted by $\mathcal{M}(\Omega)$. it is the topological dual of $C_c(\Omega)$. In order that a linear map

$$\mu: C_c(\Omega) \to \underline{C}$$

be a Radon measure it is necessary and sufficient that for every compact subset K of Ω, the linear map μ restricted to $C_c(\Omega; K)$ be continuous. This follows immediately from Proposition 1.1. Equivalently, the linear map μ is a Radon measure on Ω if and only if for every compact subset K of Ω there is a constant M_K such that

$$|\mu(\phi)| \leq M_K \cdot \sup_{x \in K} |\phi(x)|, \forall \phi \in C_c(\Omega; K).$$

As examples of Radon measures let us mention that every locally integrable function f in Ω (in particular, every continuous function) defines a Radon measure on Ω, by setting

$$\mu_f(\phi) = \int_\Omega f\phi, \forall \phi \in C_c(\Omega).$$

Furthermore, if $L^1_{\ell oc}(\Omega)$ represents the vector space of all classes of locally integrable functions on Ω, the map

$$f \to \mu_f$$

gives an imbedding of $L^1_{\ell oc}(\Omega)$ into $\mathcal{M}(\Omega)$. Indeed, if $\mu_f \equiv 0$ on $C_c(\Omega)$ then f must be zero almost everywhere in Ω, hence it defines the zero element of $L^1_{\ell oc}(\Omega)$.

The reader should consult Bourbaki [5] for a more general discussion on Radon measures.

2. *The spaces* $L^p_c(\Omega)$, $1 \leq p \leq +\infty$. Let Ω be an open subset of $\underline{R}^n$ and let K be an arbitrary compact subset of Ω. Denote by $L^p(K)$, $1 \leq p < +\infty$, the space of pth *power integrable functions with compact support contained in* K, equipped with its natural norm. When $p = \infty$, $L^\infty(K)$ is the space of *essentially bounded functions with*

compact support in K, equipped with its natural norm. The spaces $L^p(K)$, $1 \leq p \leq +\infty$, are Banach spaces; if $K_1 \subset K_2$ the imbedding $L^p(K_1) \to L^p(K_2)$ is continuous and the topology induced by $L^p(K_2)$ on $L^p(K_1)$ coincides with the topology of $L^p(K_1)$. Let $L^p_c(\Omega)$ be the union of $L^p(K)$ for all compact subsets K of Ω. If (K_i) is an increasing sequence of compact subsets of Ω whose union is Ω, it is clear that $L^p_c(\Omega)$ and $L^p(K_i)$, $i = 1, 2, \cdots$, satisfy all the assumptions of Theorem 1.3. We can define on $L^p_c(\Omega)$ the inductive limit topology relative to its subspaces $L^p(K_i)$, $i = 1, 2, \cdots$.

PROBLEMS

1. Use the Leibniz formula to expand $\Delta(u \cdot v)$ and $\Delta^2(u \cdot v)$, where

$$\Delta = \sum_{i=1}^{n} \frac{\partial^2}{\partial x_i^2} .$$

2. Write the Leibniz formula (1.1) in the case n = 1.

3. If $p \in \underline{N}^n$, show that

$$\partial^p(u \cdot v) = \sum_{q+r=p} \frac{p!}{q!r!} \partial^q u \cdot \partial^r v.$$

4. Let $\langle x, \xi \rangle = x_1\xi_1 + \cdots + x_n\xi_n$ and let $P(\partial)$ be a partial differential operator. Show that

$$P(\partial)(u(x)e^{\langle x,\xi \rangle}) = e^{\langle x,\xi \rangle} P(\partial+\xi)u(x).$$

5. Prove that the function $\beta(x)$ defined in Section 2 is a C^∞ function in $\underline{R}^n$.

6. Let v be a continuous function with compact support in $\underline{R}^n$ and let v_ε be its convolution with α_ε. Prove that: (i) $v_\varepsilon \to v$ uniformly on $\underline{R}^n$; (ii) $v_\varepsilon \to v$ in $L^p(\underline{R}^n)$, $1 \leq p + < +\infty$.

7. Let $C_0(\underline{R}^n)$ be the space of all continuous functions f in $\underline{R}^n$ converging to zero at ∞, i.e., such that given $\varepsilon > 0$, there is

a compact subset $K \subset \underline{R}^n$ such that $|f(x)| < \epsilon$ for all $x \notin K$. Prove that every $f \in C_0(\underline{R}^n)$ can be uniformly approximated in $\underline{R}^n$ by functions of $C_c^\infty(\underline{R}^n)$.

8. Prove that the maps defined in Section 3, Examples 1 and 2 are norms and the ones defined in Examples 3 and 4 are seminorms.

9. Let E be a seminormed space, i.e., a vector space equipped with a topology defined by a seminorm p. Prove that the maps

$$(x, y) \in E \times E \to x + y \in E$$

$$(\lambda, x) \in \kappa \times E \quad \lambda x \in E$$

are continuous.

10. Prove that the sets $V(x_0, \varepsilon, F)$ on p. 13 define a fundamental system of neighborhoods of x_0 of a topology compatible with the vector space structure of E.

11. Let E be a topological vector space. Prove that the maps

$$x \in E \to a + x \in E$$

and

$$x \in E \to \lambda x \in E,$$

where $a \in E$ and $\lambda \in \kappa$, $\lambda \neq 0$, are homeomorphisms on E.

12. Prove that the intersection of convex sets is a convex set.

13. Prove that the convex hull $\Gamma(A)$ of A is the set described on p. 15.

14. Let q be the seminorm defined by a balanced, convex, and absorbing set V. Prove that

$$\{x \in E: q(x) < 1\} \subset V \subset \{x \in E: q(x) \leq 1\}.$$

15. In Theorem 1.2, complete the proof that condition 3 implies condition 1.

16. Let E be a vector space and let M be a subset of E^*, the algebraic dual of E. Prove that the locally convex topology defined on E by the family of seminorms $(p_{x^*})_{x^* \in M}$ coincides with the locally convex one defined by the family of seminorms $(p_{x^*})_{x^* \in F}$, where F is the subspace of E^* spanned by M.

17. Let E be a topological vector space and let E' be its dual. Prove that the weak topology $\sigma(E', E)$ is coarser than the strong topology.

18. Let E be a topological vector space. Prove that a sequence (x_j') converges to zero in the strong dual E_b' if and only if the numerical sequence $(x_j'(x))$ converges to zero uniformly on every bounded set of E.

19. Let $\mathcal{G}$ be a family of bounded sets of a topological vector space E. Show that the topology on E' defined by the family of seminorms $(p_{A^\circ})_{A \in \mathcal{G}}$ remains the same if we replace $\mathcal{G}$ by the family $\mathcal{G}_1$ of bounded sets of E obtained by taking finite union of homothetic sets of $\mathcal{G}$.

20. Let E and F be two topological vector space and let $u: E \to F$ be a continuous linear map. Prove that if A is bounded in E, u(A) is bounded in F.

21. If A and B are convex sets, show that every $z \in \Gamma(A \cup B)$ can be written as $z = \alpha x + \beta y$ with $x \in A$, $y \in B$, $\alpha \geq 0$, $\beta \geq 0$, and $\alpha + \beta = 1$. If A and B are balanced convex sets, show that every $z \in \Gamma_b(A \cup B)$ can be written as $z = \alpha x + \beta y$, with $x \in A$, $y \in B$, and $|\alpha| + |\beta| \leq 1$.

22. Prove that the set W constructed in the proof of Lemma 1.1 is open.

23. Prove that if Ω is an open subset of $\underline{R}^n$, there is an increasing sequence (K_j) of compact sets such that $K_j \subset \Omega$ and $\bigcup_j K_j = \Omega$.

24. Let $C_c(\Omega; K)$ be equipped with the topology of uniform convergence on K. Prove that: (i) $C_c(\Omega; K)$ is a Banach space; (ii) its topology coincides with the one induced by $C(\Omega)$.

25. Let K and L be two compact subsets of Ω such that $K \subset L$. Prove that the imbedding $C_c(\Omega; K) \to C_c(\Omega; L)$ is continuous and that $C_c(\Omega; K)$ is a closed subspace of $C_c(\Omega; L)$.

26. Prove the characterization of convergent sequences in $C_c(\Omega)$ as stated in Section 6, Example 1.

27. Prove that a linear function $\mu: C_c(\Omega) \to \underline{C}$ is a Radon measure if and only if for every compact subset $K \subset \Omega$ there is a constant M_K such that

$$|\mu(\phi)| \leq M_K \cdot \sup_{x \in K} |\phi(x)|, \quad \phi \in C_c(\Omega; K).$$

28. We say that a Radon measure μ on Ω is equal to *zero* on an open subset U of Ω if and only if $\mu(\psi) = 0$ for all $\psi \in C_c(U)$. Prove that if μ is zero on U_1 and on U_2, then μ is zero on $U_1 \cup U_2$. (*Hint:* Use a partition of unity.)

29. Call the support of a Radon measure μ on Ω the complement in Ω of the largest open subset of Ω where μ is zero. Show that if $f \in L^1_{\ell oc}(\Omega)$, the support of f as a function or a Radon measure do coincide.

30. Let $(C(\Omega))'$ be the dual of $C(\Omega)$ equipped with its natural topology. Prove that $\mu \in (C(\Omega))'$ if and only if there is a constant $C > 0$ and a compact set $K \subset \Omega$ such that

$$|\mu(\phi)| \leq C \cdot \sup_{x \in K} |\phi(x)|, \quad \phi \in C(\Omega).$$

31. (i) Prove that every $\mu \in (C(\Omega))'$ defines a Radon measure on Ω.

(ii) Prove that $(C(\Omega))'$ can be identified with a subspace of $M(\Omega)$, the space of all Radon measures on Ω. [*Hint:* $C_c(\Omega)$ is a dense subspace of $C(\Omega)$.]

(iii) Prove that the elements of $(C(\Omega))'$ are Radon measures on Ω with compact support.

Chapter 2

DISTRIBUTIONS

1. THE TOPOLOGY OF $C^\infty(\Omega)$

Let Ω be an open subset of $\underline{R}^n$ and let $(K_j)_{j=1,2,\cdots}$ be an increasing sequence of compact subsets of Ω whose union is Ω. Denote by $C^m(\Omega)$ (or $\mathcal{E}^m(\Omega)$ in Schwartz's notation [28]) the space of all complex-valued functions having continuous partial derivatives of order $\leq$m. For every j = 1, 2,$\cdots$ define the seminorm

$$p_{m,j}(\phi) = \sup_{\substack{x\in K_j \\ |\alpha|\leq m}} |\partial^\alpha\phi(x)|, \forall\ \phi \in C^m(\Omega).$$

The family of seminorms $(p_{m,j})_{j=1,2,\cdots}$ defines a Hausdorff locally convex topology on $C^m(\Omega)$. Since it is a countable family, every element of $C^m(\Omega)$ has a countable fundamental system of neighborhoods, hence $C^m(\Omega)$ is a metrizable space.

The topology of $C^m(\Omega)$, which we often call the natural topology of $C^m(\Omega)$, is the *topology of uniform convergence on compact subsets of Ω of the functions and their derivatives of order* $\leq$m. Indeed, we have the following result.

Theorem 2.1. *A necessary and sufficient condition for a sequence* (ϕ_k) *to converge to zero in* $C^m(\Omega)$ *is that, for every* $|\alpha| \leq m$, *the sequence* $(\partial^\alpha\phi_k)$ *converge uniformly on every compact subset of* Ω.

Proof. Let $|\alpha| \leq m$ and let K be a compact subset of Ω. Choose j so that $K_j \supset K$ and let V be the following neighborhood of zero in

$C^m(\Omega)$:

$$V = \{\phi \in C^m(\Omega): p_{m,j}(\phi) \leq \varepsilon\}$$

where ε is a positive real number. If the sequence (ϕ_k) converges to zero in $C^m(\Omega)$, we can find an index k_0 such that $\phi_k \in V$, $k \geq k_0$. In other words,

$$\sup_{\substack{x \in K_j \\ |\alpha| \leq m}} |\partial^\alpha \phi_k(x)| \leq \varepsilon, \forall k \geq k_0,$$

which means, precisely, that the sequence $(\partial^\alpha \phi_k)$ converges to zero uniformly on K_j, hence on K. As an exercise, we leave to the reader the proof of the sufficient condition.

It is easily seen that the natural topology of $C^m(\Omega)$ is the coarsest one for which the linear maps

$$\partial^\alpha: C^m(\Omega) \to C(\Omega), \quad |\alpha| \leq m,$$

are continuous, the space $C(\Omega)$ being equipped with its natural topology (see Problem 5). We can then rephrase Theorem 2.1 as follows: *A sequence* (ϕ_k) *converges to zero in* $C^m(\Omega)$ *if and only if the sequence* $(\partial^\alpha \phi_k)$ *converges to zero in* $C(\Omega)$ *for all* $|\alpha| \leq m$.

Theorem 2.2 *The locally convex space* $C^m(\Omega)$ *is complete.*

Proof. Let (ϕ_k) be a Cauchy sequence in $C^m(\Omega)$. In particular, (ϕ_k) is a Cauchy sequence in $C(\Omega)$. But $C(\Omega)$ is a complete space (Chapter 1, Section 4, Example 2); hence, the sequence (ϕ_k) converges to an element $\psi \in C(\Omega)$. On the other hand, for every $1 \leq i \leq n$, the sequence $(\partial_i \phi_k)$ is also a Cauchy sequence in $C(\Omega)$. Hence, $\partial_i \phi_k \to \psi_i$ in $C(\Omega)$, $1 \leq i \leq n$. From a classical theorem in analysis [26, p. 140] it follows that $\psi_i = \partial_i \psi$, $1 \leq i \leq n$. Proceeding by induction, we can prove that for every $|\alpha| \leq m$ the sequence $(\partial^\alpha \phi_k)$ converges to an element ψ_α in $C(\Omega)$ such that $\psi_\alpha = \partial^\alpha \psi$. Q.E.D.

According to Definition 1.12, $C^m(\Omega)$ is then a Frechet space.

Let $C^\infty(\Omega)$ (or $E(\Omega)$ in Schwartz's notation [28]) be the space of infinitely differentiable functions in Ω. The countable family of seminorms $(p_{m,j})_{m=0,1,2,\cdots;\ j=1,2,3,\cdots}$ defines a locally convex Hausdorff topology on $C^\infty(\Omega)$ called the natural topology of $C^\infty(\Omega)$. It is easy to see that Theorems 2.1 and 2.2 are true on replacing $C^m(\Omega)$ by $C^\infty(\Omega)$. Hence, $C^\infty(\Omega)$ is a Frechet space whose topology coincides with the coarsest one for which the linear maps

$$\partial^\alpha : C^\infty(\Omega) \to C(\Omega),\ \forall\ \alpha,$$

are continuous, the space $C(\Omega)$ being equipped with its natural topology. Also, the topology of $C^\infty(\Omega)$ is the coarsest one for which the identity map $C^\infty(\Omega) \to C^m(\Omega)$ is continuous for all $m \geq 0$.

The next theorem characterizes *bounded* sets in $C^\infty(\Omega)$.

Theorem 2.3. *A subset* A *of* $C^\infty(\Omega)$ *is bounded if and only if, given an integer* $m \geq 0$ *and a compact set* $K \subset \Omega$ *there is a constant* $C > 0$ *such that*

$$|\partial^\alpha \phi(x)| \leq C, \forall |\alpha| \leq m, \forall\ x \in K, \forall\ \phi \in A.$$

Proof. Suppose that the set A satisfies this condition. In order to show that A is bounded, we must prove that, given V a neighborhood of zero in $C^\infty(\Omega)$, we can find a number $\lambda > 0$ such that $\lambda A \subset V$. We can obviously assume that the neighborhood V is of the following form: $V = \{\phi \in C^\infty(\Omega) : p_{m,j}(\phi) \leq \varepsilon\}$. By our assumption on A, there is a constant $C > 0$ such that

$$|\partial^\alpha \phi(x)| \leq C, \forall\ |\alpha| \leq m, \forall\ x \in K_j, \forall\ \phi \in A.$$

Let λ be a positive real number such that $\lambda C \leq \varepsilon$. Then the last inequality implies that

$$\lambda |\partial^\alpha \phi(x)| \leq \varepsilon,\ \forall |\alpha| \leq m,\ \forall x \in K_j,\ \forall\ \phi \in A.$$

In other words, $\lambda A \subset V$, hence A is bounded in $C^\infty(\Omega)$. As an exercise, we leave to the reader the proof that if A is a bounded subset of $C^\infty(\Omega)$ then the condition of our theorem is satisfied.

Relatively Compact Subsets of $C^\infty(\Omega)$

We recall that a subset of a topological space is *relatively compact* if its closure is compact. In what follows, we shall give a characterization of relatively compact subsets of $C^m(\Omega)$ based on *Ascoli's theorem* and we shall derive, as a consequence, the fact that in $C^\infty(\Omega)$ relatively compact and bounded subsets do coincide.

Definition 2.1. *A subset* A *of* $C(\Omega)$ *is said to be equicontinuous at a point* $x_0 \in \Omega$ *if, given* $\varepsilon > 0$ *there is neighborhood* U *of* x_0 *such that*

$$|\phi(x) - \phi(x_0)| \leq \varepsilon, \forall x \in U, \forall \phi \in A.$$

A *is equicontinuous in* Ω *if it is equicontinuous at every point of* Ω.

The following is one of several versions of Ascoli's theorem (see [4, 18]).

Theorem 2.4 (Ascoli). *A subset* A *of* $C(\Omega)$ *is relatively compact if and only if* (1) A *is equicontinuous;* (2) *for every* $x \in \Omega$ *the subset* $\{\phi(x):\phi \in A\}$ *is relatively compact in* $\underline{C}$.

Theorem 2.5 *A subset* A *of* $C^m(\Omega)$ *is relatively compact if and only if* (1) A *is bounded in* $C^m(\Omega)$; (2) *for every* $p = (p_1,\cdots,p_n)$ *such that* $|p| = m$ *the set* $\partial^p A$ (image of A by the differential operator ∂^p) *is equicontinuous.*

The proof of Theorem 2.5 is based upon the following result.

Lemma 2.1. *Let* B *be a bounded subset in* $C^1(\Omega)$. *Then* B *is an equicontinuous subset of* $C(\Omega)$.

Proof. By the mean value theorem [7,p. 268] we have

$$|\phi(x+h) - \phi(x)| \leq \sup_{0\leq\theta\leq 1} \| \phi'(x+\theta h) \| \cdot |h|$$

where $h = (h_1, \cdots, h_n)$, ϕ' denotes the *differential* of ϕ, and $\|\phi'\|$ denotes its norm as a linear map from $\underline{R}^n$ into $\underline{C}$.

Since B is bounded in $C^1(\Omega)$ it is easy to see that there is a constant $M > 0$ such that

$$\sup_{0\leq\theta\leq 1} \| \phi'(x+\theta h)\| \leq M, \ \forall \ \phi \in B.$$

Replacing in the above we get

$$|\phi(x+h) - \phi(x)| \leq M \cdot |h| \ , \forall \ \phi \in B,$$

hence B is an equicontinuous set. Q.E.D.

Proof of Theorem 2.5. Suppose that A is a relatively compact subset of $C^m(\Omega)$. Then A is a bounded subset of $C^m(\Omega)$ (Chapter 1, Section 5, Example 3). On the other hand, by Ascoli's theorem it follows that $\partial^p A$ is equicontinuous for every $|p| = m$.

Conversely, suppose that conditions 1 and 2 hold true. Let $p = (p_1, \cdots, p_n)$ be such that $|p| = m$ and consider the set $\partial^p A$. By assumption, $\partial^p A$ is bounded and equicontinuous; hence, by Ascoli's theorem, $\partial^p A$ is a relatively compact set. Now let $q = (q_1, \cdots, q_n)$ be such that that $|q| = m-1$ and let $B = \partial^q A$. Obviously, B is a bounded subset of $C^1(\Omega)$; hence, by Lemma 2.1, B is an equicontinuous set. Again by Ascoli's theorem, $B = \partial^q A$ is a relatively compact set. An induction argument shows that $\partial^p A$ is a relatively compact subset of $C(\Omega)$ for all $|p| \leq m$; hence by Theorem 2.4, A is a relatively compact subset of $C^m(\Omega)$. Q.E.D.

Corollary. Every bounded subset A *of* $C^{m+1}(\Omega)$ *is relatively compact in* $C^m(\Omega)$.

Proof. It suffices to apply the lemma and the previous theorem. Q.E.D.

As an easy consequence of the preceding results we have the following.

Theorem 2.6. *A subset* A *of* $C^\infty(\Omega)$ *is relatively compact if and only if* A *is bounded in* $C^\infty(\Omega)$.

Following Grothendieck [12], we state the following definition.

Definition 2.2. *We say that a Hausdorff locally convex space* E *is a Montel space if every bounded subset of* E *is relatively compact.*

The space $C^\infty(\Omega)$ equipped with its natural topology is then a Montel space. Let us point out that the above definition differs from that of Bourbaki [6], Horvath [17], or Treves [32]. However, in all locally convex spaces defined here the two notions do coincide.

Every Montel space is a reflexive space [6, 17]. Hence, $C^\infty(\Omega)$ is a reflexive space.

The inductive limit of Montel spaces is a Montel space. Indeed, let E be an inductive limit of a sequence $(E_i)_{i=1,2,\cdots}$ (Chapter 1, Section 6) and suppose that E_i is a Montel space. Let A be a bounded set in E. By Theorem 1.3, there is an index i_0 such that $A \subset E_{i_0}$ and A is bounded in E_{i_0}. Since E_{i_0} is a Montel space, then A is relatively compact in E_{i_0}, hence *in* E, because the identity map from E_{i_0} into E is continuous; therefore, by Definition 2.2, E is a Montel space.

2. THE TOPOLOGY OF $C_c^\infty(\Omega)$

Let K be a compact subset of Ω and denote by $C_c^\infty(\Omega;\ K)$ the subspace of $C_c^\infty(\Omega)$ consisting of all functions having support contained in K. On $C_c^\infty(\Omega;\ K)$ we consider the topology induced by $C^\infty(\Omega)$ that coincides with the locally convex topology defined by the sequence of norms

$$p_m(\phi) = \sup_{\substack{x \in K \\ |\alpha| \leq m}} |\partial^\alpha \phi(x)|, \quad m \in \underline{N}.$$

Theorem 2.7. $C_c^\infty(\Omega;\ K)$ *is a Frechet space.*

Proof. Since the topology of $C_c^\infty(\Omega; K)$ is defined by a sequence of norms, $C_c^\infty(\Omega, K)$ is a metrizable space. In order to show that $C_c^\infty(\Omega; K)$ is complete, it suffices to show that $C_c^\infty(\Omega; K)$ is a closed subspace of $C^\infty(\Omega)$, which is obvious. Q.E.D.

Theorem 2.8. *Let* K *and* L *be two compact subsets such that* $K \subset L \subset \Omega$. *We have:* (1) *The identity map* $C_c^\infty(\Omega; K) \to C_c^\infty(\Omega; L)$ *is a continuous one;* (2) $C_c^\infty(\Omega; K)$ *is a closed subspace of* $C_c^\infty(\Omega; L)$.

The proof is left to the reader as an exercise.

Suppose now that (K_j) is an increasing sequence of compact sets such that $K_j \subset \Omega$ and the union of K_j is Ω. It is clear that

$$C_c^\infty(\Omega) = \bigcup_j C_c^\infty(\Omega; K_j).$$

By Theorems 2.7 and 2.8 we see that the sequence of Frechet spaces $C_c^\infty(\Omega; K_j)$ satisfies all the assumptions of Theorem 1.3. We then define on $C_c^\infty(\Omega)$ the *inductive limit topology* of the spaces $C_c^\infty(\Omega; K_j)$. As a consequence of Theorem 1.3, it follows that on every subspace $C_c^\infty(\Omega; K_j)$ the topology induced by the inductive limit coincides with the topology induced by $C^\infty(\Omega)$. One can also prove that if we replace the sequence (K_j) by another sequence of compact subsets with the same properties, the inductive limit topology of $C_c^\infty(\Omega)$ remains unchanged.

Also by Theorem 1.3, in order that *a subset* A *be bounded in* $C_c^\infty(\Omega)$ *it is necessary and sufficient that there exist an index* j *such that* A *be contained and bounded in* $C_c^\infty(\Omega; K_j)$. This is equivalent to saying that a set A *is bounded in* $C_c^\infty(\Omega)$ *if and only if there is a compact subset* $K \subset \Omega$ *such that:* (i) *the support of every* $\phi \in A$ *is contained in* K; (ii) *to every* $\alpha = (\alpha_1, \cdots, \alpha_n)$ *there corresponds a constant* C_α *such that*

$$|\partial^\alpha \phi(x)| \leq C_\alpha, \ \forall\ x \in K, \ \forall\ \psi \in A.$$

We have the following criterion of convergence.

Theorem 2.9. *A sequence* (ϕ_k) *converges to zero in* $C_c^\infty(\Omega)$ *if and only if we can find a compact subset* $K \subset \Omega$ *such that:* (1) *the support of every function* ϕ_k *is contained in* K; (2) *for every* α *the sequence* $(\partial^\alpha \phi_k)$ *converges to zero uniformly on* K.

Proof. Suppose that $(\phi_k) \to 0$ as $k \to +\infty$. Then the sequence (ϕ_k) is a bounded subset of $C_c^\infty(\Omega)$. By Theorem 1.3, there is a compact subset $K \subset \Omega$ such that $\phi_k \in C_c^\infty(\Omega; K)$ for all k. On the other hand, since the topology of $C_c^\infty(\Omega; K)$ coincides with the one induced by $C_c^\infty(\Omega)$, it follows that $(\phi_k) \to 0$ in $C_c^\infty(\Omega; K)$.

Conversely, the conjunction of 1 and 2 implies that $(\phi_k) \to 0$ in $C_c^\infty(\Omega; K)$. Since the identity map $C_c^\infty(\Omega; K_j) \to C_c^\infty(\Omega)$ is continuous, the given sequence converges in $C_c^\infty(\Omega)$. Q.E.D.

Theorem 2.10. *The identity map* $C_c^\infty(\Omega) \to C^\infty(\Omega)$ *is continuous.*

Proof. By definition, the imbedding $C_c^\infty(\Omega; K_j) \to C^\infty(\Omega)$ is continuous for every j. It then follows from Proposition 1.1 the continuity of the identity map $C_c^\infty(\Omega) \to C^\infty(\Omega)$. Q.E.D.

As a consequence of Theorem 2.10, every bounded subset of $C_c^\infty(\Omega)$ is a bounded subset of $C^\infty(\Omega)$.

Theorem 2.11. $C_c^\infty(\Omega)$ *is a Montel space.*

Proof. Indeed, $C_c^\infty(\Omega)$ is an inductive limit of a sequence of Montel spaces. Q.E.D.

As a consequence, we have the following result.

Corollary. $C_c^\infty(\Omega)$ *is a reflexive space.*

Let us also remark that $C_c^\infty(\Omega)$ is a *complete space*. Since $C_c^\infty(\Omega)$ is *not* a metrizable space, one has to show that every *Cauchy filter* converges in $C_c^\infty(\Omega)$. We shall not prove this result here. The reader should consult for instance, Horvath [17, p. 162-165], where a more general case is discussed.

However, it is quite simple to see that $C_c^\infty(\Omega)$ is *sequentially complete*. Indeed, if (ϕ_j) is a Cauchy sequence in $C_c^\infty(\Omega)$, it is bounded; hence, by Theorem 1.3, it is a Cauchy sequence in some space $C_c^\infty(\Omega; K)$; hence, it converges in $C_c^\infty(\Omega; K)$ because this space is complete (Theorem 2.7). Therefore, (ϕ_j) converges in $C_c^\infty(\Omega)$.

Complements

Let $C_c^m(\Omega)$ be the space of all functions of class C^m having compact support contained in Ω and let $C_c^m(\Omega; K)$ be the subspace of $C^m(\Omega)$ consisting of all functions with compact support contained in K, a compact subset of Ω. On $C_c^m(\Omega; K)$ we define the topology induced by $C^m(\Omega)$ which coincides with the topology defined by the norm

$$p_m(\phi) = \sup_{\substack{x \in K \\ |\alpha| \leq m}} |\partial^\alpha \phi(x)|.$$

It is easy to verify that $C_c^m(\Omega; K)$ is a Banach space.

As is the case of the space $C_c^\infty(\Omega)$, we define on $C_c^m(\Omega)$ the inductive limit topology of a sequence of subspaces $C_c^m(\Omega; K_j)$, where (K_j) is an increasing sequence of compact subsets of Ω whose union is Ω. The characterizations of convergent sequences and of bounded sets in the space $C_c^m(\Omega)$ are analogous to those in $C_c^\infty(\Omega)$. However, $C_c^m(\Omega)$ it is *not* a reflexive space.

It is easy to see that the imbedding $C_c^\infty(\Omega) \to C_c^m(\Omega)$ is continuous for all $m \geq 0$. We can then define on $C_c^\infty(\Omega)$ the *coarsest* topology for which these imbeddings are continuous and, for reasons that shall appear later, we denote by $\mathcal{D}_F(\Omega)$ the space $C_c^\infty(\Omega)$ equipped with such topology. As we shall see in Section 7, the natural topology of $C_c^\infty(\Omega)$ is *strictly finer* than the topology of $\mathcal{D}_F(\Omega)$.

3. DISTRIBUTIONS

Definition 2.3. Let Ω *be an open subset of* $\underline{R}^n$. *A distribution on* Ω *is a continuous linear functional on* $C_c^\infty(\Omega)$.

The vector space of all distributions on Ω will be denoted by $\mathcal{D}'(\Omega)$. It is then the topological dual of $C_c^\infty(\Omega)$.

Examples. 1. Let f be a locally integrable function on Ω. Define the linear functional

$$T_f\colon C_c^\infty(\Omega) \to \underline{C}$$

by means of

$$T_f(\phi) = \int_\Omega f\cdot\phi, \ \forall\,\phi \in C_c^\infty(\Omega).$$

As we know (Section 6, Example 1), T_f defines a Radon measure on Ω, i.e., a continuous linear functional on $C_c(\Omega)$. On the other hand, the imbedding

$$C_c^\infty(\Omega) \to C_c(\Omega)$$

is a continuous one, hence T_f defines a continuous linear functional on $C_c^\infty(\Omega)$, i.e., a distribution on Ω. When no confusion is possible, we shall denote T_f simply by f.

In particular, functions of $C^m(\Omega)$, $0 \le m \le +\infty$, L^p functions, $1 \le p \le +\infty$, define distributions on Ω.

2. More generally, every Radon measure μ on Ω defines a distribution on Ω. Moreover, we can identity $M(\Omega)$ with a subspace of $\mathcal{D}'(\Omega)$:

$$M(\Omega) \subset \mathcal{D}'(\Omega).$$

Indeed, it suffices to show that if $\mu \in M(\Omega)$ is such that

$$\mu(\phi) = 0, \forall\phi \in C_c^\infty(\Omega),$$

then μ is the Radon measure identically zero, i.e., $\mu(\phi) = 0$. $\forall\, \psi \in C_c(\Omega)$. If ψ is an element of $C_c(\Omega)$, then by Theorem 1.1,

$$\psi_\varepsilon = \alpha_\varepsilon * \psi$$

is an element of $C_c^\infty(\Omega)$ and it is easy to see that $\psi_\varepsilon \to \psi$ *in* $C_c(\Omega)$ when $\varepsilon \to 0$. Therefore, by continuity, $\mu(\psi) = 0$. Q.E.D.

3. The *Dirac measure* δ on $\underline{R}^n$ defined by

$$\delta(\phi) = \phi(0), \forall\, \phi \in C_c^\infty(\underline{R}^n)$$

is a distribution.

4. The following is an example of a distribution which is *not* a measure. Let $\Omega = \underline{R}$ and define

$$\delta'(\phi) = \delta\left(-\frac{d\phi}{dx}\right) = -\frac{d\phi}{dx}(0), \forall\, \phi \in C_c^\infty(\underline{R}).$$

It is clear that δ' defines a continuous linear functional on $C_c^\infty(\underline{R})$ (and even on $C_c^1(\underline{R})$) but *not* on $C_c^0(\underline{R})$. This example can be extended to the case of more variables. Let $\Omega = \underline{R}^n$ and set

$$<\partial_k\delta, \phi> = -<\delta, \partial_k\phi> = -\partial_k\phi(0)$$

for all $\phi \in C_c^\infty(\underline{R}^n)$, where $1 \leq k \leq n$. Later on we shall see that $\partial_k\delta$ is precisely the partial derivative of the Dirac measure in the *sense of distributions*.

5. The function of one variable $1/x$ is *not* locally integrable on $\underline{R}^n$, hence it does not define a distribution on $\underline{R}$, in the sense of Example 1. By definition, we set

$$<PV\frac{1}{x}, \phi> = PV \int_{-\infty}^{+\infty} \frac{\phi(x)}{x}\,dx = \lim_{\varepsilon\to 0} \int_{|x|\geq\varepsilon} \frac{\phi(x)}{x}\,dx, \forall \phi \in C_c^{\infty}(\underline{R}),$$

where the limit on the right-hand side is called the *Cauchy principal value* of the integral $\int_{-\infty}^{+\infty} [\phi(x)/x]\,dx$. We have

$$\int_{|x|\geq\varepsilon} \frac{\phi(x)}{x}\,dx = \int_{-\infty}^{-\varepsilon} \frac{\phi(x)}{x}\,dx + \int_{\varepsilon}^{+\infty} \frac{\phi(x)}{x}\,dx$$

$$= \phi(-\varepsilon)\ \ell og\ \varepsilon - \int_{-\infty}^{-\varepsilon} \phi'(x)\ \ell og|x|\,dx - \phi(\varepsilon)\ \ell og\ \varepsilon$$

$$- \int_{\varepsilon}^{+\infty} \phi'(x)\ \ell og|x|\,dx.$$

Writing $\phi(x) = \phi(0) + x\psi(x)$ with $\psi(0) = \phi'(0)$ and replacing above, we get

$$\int_{|x|\geq\varepsilon} \frac{\phi(x)}{x}\,dx = -\ 2\varepsilon\psi(\varepsilon)\ \ell og\ \varepsilon - \int_{-\infty}^{-\varepsilon} \phi'(x)\ \ell og|x|\,dx$$

$$- \int_{\varepsilon}^{+\infty} \phi'(x)\ \ell og|x|\,dx.$$

Taking limits, we obtain

$$\lim_{\varepsilon\to 0} \int_{|x|\geq\varepsilon} \frac{\phi(x)}{x}\,dx = -\int_{-\infty}^{+\infty} \phi'(x)\ \ell og|x|\,dx,$$

where the last integral converges and defines a continuous linear functional on $C_c^{\infty}(\underline{R})$. Hence,

$$<PV\frac{1}{x}, \phi> = \lim_{\varepsilon\to 0} \int_{|x|\geq\varepsilon} \frac{\phi(x)}{x}\, dx = -\int_{-\infty}^{+\infty} \phi'(x)\, \ell og|x|\, dx$$

defines a distribution on $\underline{R}$.

From Definition 2.3 and Proposition 1.1 it follows that T is a distribution on Ω if and only if for every compact subset K contained in Ω, T is a continuous linear functional on $C_c^\infty(\Omega; K)$. In other words, the following holds.

Theorem 2.12. A linear functional T *is a distribution on* Ω *if and only if for every compact set* $K \subset \Omega$ *there is a constant* $C > 0$ *and an integer* $m \geq 0$ *such that*

$$|<T, \phi>| \leq C\cdot \sup_{\substack{x\varepsilon\Omega \\ |\alpha|\leq m}} |\partial^\alpha \phi(x)|, \forall \psi \in C_c^\infty(\Omega; K). \tag{2.1}$$

Proof. Let $T \in \mathcal{D}'(\Omega)$. As we remarked above, for every compact subset $K \subset \Omega$, T is a continuous linear functional on $C_c^\infty(\Omega; K)$. Then, there is a neighborhood of zero

$$V = V(K, m, \varepsilon) = \{\phi \in C_c^\infty(\Omega; K): p_{m,K}(\phi) \leq \varepsilon\}$$

where

$$p_{m,K}(\phi) = \sup_{\substack{x\in K \\ |\alpha|\leq m}} |\partial^\alpha \phi(x)|,$$

such that

$$|<T, \phi>| \leq 1, \forall \phi \in V.$$

On the other hand, if $\phi \in C_c^\infty(\Omega;\ K)$ is such that $\phi \neq 0$ we have

$$\frac{\varepsilon\phi}{p_{m,K}(\phi)} \in V.$$

It follows that

$$|\langle T,\ \phi\rangle| < \frac{1}{\varepsilon}\, p_{m,K}(\phi),\ \forall\, \phi \in C_c^\infty(\Omega;\ K),\ \phi \neq 0.$$

By setting $C = \varepsilon^{-1}$ we get (2.1) for all $\phi \in C_c^\infty(\Omega;\ K)$ with $\phi \neq 0$. Finally, notice that (2.1) is an identity when $\phi = 0$.

Conversely, if (2.1) is satisfied, it implies that for every $j = 1, 2, \cdots$, T is a continuous linear functional on $C_c^\infty(\Omega;\ K_j)$. Therefore (Proposition 1.1) T is a continuous linear functional on $C_c^\infty(\Omega)$. Q.E.D.

Let us point out that condition (2.1) is not a simple one to verify. When proving whether or not a linear functional T is a distribution, we shall often use the following characterization of distributions in terms of *convergent sequences* in $C_c^\infty(\Omega)$.

Theorem 2.13. $T \in \mathcal{D}'(\Omega)$ *if and only if for every sequence* (ϕ_k) *converging to zero in* $C_c^\infty(\Omega)$ *the numerical sequence* $(\langle T,\phi_k\rangle)$ *converges to zero.*

Proof. Suppose that $T \in \mathcal{D}'(\Omega)$ and let (ϕ_k) be a sequence converging to zero in $C_c^\infty(\Omega)$. By Theorem 2.9, there is a compact subset $K \subset \Omega$ such that

$$\phi_k \in C_c^\infty(\Omega;\ K),\ \forall k,$$

and

$$\phi_k \to 0 \text{ in } C_c^\infty(\Omega;\ K).$$

Since T is continuous on $C_c^\infty(\Omega; K)$ (Proposition 1.1),

$$(\langle T, \phi_k \rangle) \to 0 \text{ as } k \to +\infty.$$

Conversely, suppose that this condition is satisfied. In order to prove that T, a linear functional on $C_c^\infty(\Omega)$, is continuous, it suffices to show that T is continuous on every subspace

$$C_c^\infty(\Omega; K_j),\ j = 1, 2, \cdots$$

(Proposition 1.1). Let us suppose, by contradiction, that there is an index j_0 such that T is *not* continuous on $C_c^\infty(\Omega; K_{j_0})$. Then, we can find a sequence (ϕ_k) of functions of $C_c^\infty(\Omega; K_{j_0})$ converging to zero in $C_c^\infty(\Omega; K_{j_0})$ but such that the sequence $(\langle T, \phi_k \rangle)$ does not converge. Since the imbedding

$$C_c^\infty(\Omega; K_{j_0}) \to C_c^\infty(\Omega)$$

is continuous, (ϕ_k) must converge to zero in $C_c^\infty(\Omega)$, hence $(\langle T, \phi_k \rangle)$ must converge to zero, which is a contradiction. Q.E.D.

Topologies on $\mathcal{D}'(\Omega)$

Among the topologies on $\mathcal{D}'(\Omega)$ compatible with the vector space structure, the most important are the *weak topology* and the *strong topology*. In Chapter 1, Section 5, we have discussed the definition of these topologies in the general case of a topological vector space. From our discussion it follows that the weak topology on $\mathcal{D}'(\Omega)$ is the locally convex one defined by the family of seminorms associated with the elements of $C_c^\infty(\Omega)$, namely

$$p_\phi(T) = |\langle T, \phi\rangle|$$

with $\phi \in C_c^\infty(\Omega)$ and $T \in \mathcal{D}'(\Omega)$. We have the following criterion of convergence for *sequences* of distributions:

A sequence (T_j) *of elements of* $\mathcal{D}'(\Omega)$ *converges weakly to zero if and only if for every* $\phi \in C_c^\infty(\Omega)$ *the numerical sequence* $(\langle T_j, \phi\rangle)$ *converges to zero.*

A similar convergence criterion holds true for *filters*. The weak topology is also called the *topology of pointwise convergence* in $C_c^\infty(\Omega)$.

On the other hand, the strong topology on $\mathcal{D}'(\Omega)$ is the locally convex one defined by the family of seminorms associated with the polar sets of bounded sets of $C_c^\infty(\Omega)$ (Chapter 1, Section 5). We have the following convergence criterion:

A sequence (T_j) *of elements of* $\mathcal{D}'(\Omega)$ *converges strongly to zero if and only if the numerical sequence* $(\langle T_j, \phi\rangle)$ *converges to zero uniformly on every bounded subset of* $C_c^\infty(\Omega)$.

We just mention that a similar convergence criterion can be proved for *filters* of distributions. The strong topology on $\mathcal{D}'(\Omega)$ is that of *uniform convergence on bounded sets of* $C_c^\infty(\Omega)$. It is quite clear that strong convergence implies weak convergence.

The Dual of $C^{\infty}(\Omega)$

Let us denote by $\mathcal{E}'(\Omega)$ the topological dual of $C^{\infty}(\Omega)$ equipped with its natural topology. Since, by Theorem 2.10, the identity map $C_c^{\infty}(\Omega) \to C^{\infty}(\Omega)$ is continuous, every element T of $\mathcal{E}'(\Omega)$ defines a distribution on Ω. As we shall see later, the elements of $\mathcal{E}'(\Omega)$ are the *distributions with compact support on* Ω.

Theorem 2.14. $T \in \mathcal{E}'(\Omega)$ *if and only if there is a constant* $C > 0$, *an integer* $m \geq 0$, *and a compact subset* $K \subset \Omega$ *such that*

$$|\langle T, \phi\rangle| \leq C \cdot \sup_{\substack{|\alpha| \leq m \\ x \in K}} |\partial^{\alpha}\phi(x)|, \ \forall\, \phi \in C^{\infty}(\Omega). \tag{2.2}$$

Proof. If $T \in \mathcal{E}'(\Omega)$ there is a neighborhood of zero in $C^{\infty}(\Omega)$, $V = \{\phi \in C^{\infty}(\Omega) : p_{m,K}(\phi) \leq \varepsilon\}$ such that

$$|\langle T, \phi\rangle| \leq 1, \ \forall\, \phi \in V. \tag{2.3}$$

Let $\phi \in C^{\infty}(\Omega)$ be such that $p_{m,K}(\phi) \neq 0$. Then, $\varepsilon\phi/p_{m,K}(\phi) \in V$; hence

$$|\langle T, \phi\rangle| \leq C \cdot p_{m,K}(\phi),$$

with $C = \varepsilon^{-1}$. Next, if $\phi \in C^{\infty}(\Omega)$ is such that $p_{m,K}(\phi) = 0$, we claim that $\langle T, \phi\rangle = 0$. In fact, such a ϕ belongs to V, as well as any constant multiple $\lambda\phi$ of ϕ. If $\langle T, \phi\rangle$ were different from zero, then $|\langle T, \lambda\phi|$ would become as large as we please, which would contradict (2.3). Consequently, (2.2) holds true for all $\phi \in C^{\infty}(\Omega)$. We leave to the reader the proof that if (2.2) holds, then T defines a continuous linear functional on $C^{\infty}(\Omega)$. Q.E.D.

Topologies on $\mathcal{E}'(\Omega)$

As in the case of $\mathcal{D}'(\Omega)$, we shall only consider the weak and the strong topologies on $\mathcal{E}'(\Omega)$. Concerning the strong topology on $\mathcal{E}'(\Omega)$, let us prove the following.

Theorem 2.15 *A sequence* (T_j) *converges strongly to zero in* $E'(\Omega)$ *if and only if the numerical sequence* $(\langle T_j, \phi\rangle)$ *converges to zero uniformly on bounded sets of* $C^\infty(\Omega)$.

Proof. Suppose that a sequence (T_j) converges strongly to zero in $E'(\Omega)$. Let A be a bounded subset of $C^\infty(\Omega)$ and let $\varepsilon > 0$ be a given number. The set

$$V = V(A, \varepsilon) = \{T \in E'(\Omega): |p_{A^\circ}(T)| \leq \varepsilon\}$$

$$= \{T \in E'(\Omega): |\langle T, \phi\rangle| \leq \varepsilon \quad \forall \phi \in A\}$$

is a neighborhood of zero in the strong topology of $E'(\Omega)$ (Chapter 1, Section 5). Since, by assumption, (T_j) converges strongly to zero in $E'(\Omega)$, there is an index j_0 such that $T_j \in V$, $j \geq j_0$, i.e.,

$$|\langle T_j, \phi\rangle| \leq \varepsilon, \forall j \geq j_0, \forall \phi \in A,$$

which proves our assertion.

Conversely, it is easy to see that every neighborhood of zero in the strong topology of $E'(\Omega)$ contains a neighborhood of zero of the form

$$V(A_1, \cdots, A_k; \varepsilon) = \{T \in E'(\Omega); |\langle T, \phi\rangle| \leq \varepsilon, \forall\phi \in A_1 \cup \cdots \cup A_k\},$$

where $A_1, \cdots, A_k$ are bounded sets in $C^\infty(\Omega)$. Let $A = A_1 \cup \cdots \cup A_k$. The set A is bounded and since, by assumption, $\langle T_j, \phi\rangle \to 0$ uniformly when $\phi \in A$, we can find an index j_0 such that

$$|\langle T_j, \phi\rangle| \leq \varepsilon, \quad j \geq j_0, \forall\phi \in A.$$

But this relation obviously implies

$$T_j \in V(A_1, \cdots, A_k; \varepsilon), \forall\ j \geq j_0,$$

which proves that $T_j \to 0$ strongly. Q.E.D.

Examples. 1. Every locally integrable function with compact support defines by

$$<f, \phi> = \int_\Omega f \cdot \phi, \forall \phi \in C^\infty(\Omega),$$

a continuous linear functional on $C^\infty(\Omega)$. To prove the continuity it suffices to show that the map

$$\phi \to <f, \phi>$$

is already continuous on $C(\Omega)$.

If K is a compact subset of Ω and if $L^1_K(\Omega)$ denotes the space of all integrable functions with compact support contained in K, then it can be shown that

$$L^1_K(\Omega) \subset E'(\Omega).$$

In particular, $C^m_c(\Omega)$ $(m \geq 0)$ are subspaces of $E'(\Omega)$.

2. More generally, every element of $C'(\Omega)$ defines a continuous linear functional on $C^\infty(\Omega)$. Furthermore, we can identity $C'(\Omega)$ with a subspace of $E'(\Omega)$. Indeed, it suffices to show that

$$C^\infty(\Omega) \subset C(\Omega)$$

with continuous imbedding and that $C^\infty(\Omega)$ is *dense* in $C(\Omega)$.

3. The distributions δ and δ' belong to $E'(\underline{R}^n)$.

4. SUPPORT OF A DISTRIBUTION

Definition 2.4. *Let* V *be an open subset of* Ω. *We say that a distribution* $T \in \mathcal{D}'(\Omega)$ *is zero on* V *if*

$$\langle T, \phi\rangle = 0, \forall \phi \in C_c^\infty(V).$$

Examples. 1. Let $\Omega = \underline{R}$ and consider the *Heaviside function*

$$Y(x) = \begin{cases} 1 \text{ if } x \geq 0 \\ 0 \text{ if } x < 0. \end{cases}$$

Since Y is a locally integrable function, it defines a distribution on $\underline{R}$ by setting

$$\langle Y, \phi\rangle = \int_0^{+\infty} \phi, \forall \phi \in C_c^\infty(\underline{R}).$$

It is easy to see that this distribution is zero on the interval $(-\infty, 0)$. More generally, if a function f is zero in an open set, the distribution defined by f is zero on the same open set.

2. The Dirac measure δ as well as its derivatives $\partial_k \delta$, $1 \leq k \leq n$, are zero on $\underline{R}^n - \{0\}$.

Definition 2.5. *We shall say that two distributions* $S, T \in \mathcal{D}'(\Omega)$ *coincide on an open subset* V *of* Ω *if* S - T *is equal to zero on* V. *We shall write*

$$S = T \text{ on } V.$$

Lemma 2.2. *Let* $(V_i)_{i \in I}$ *be a family of open subsets of* Ω *and suppose that* $T \in \mathcal{D}'(\Omega)$ *is zero on every subset* V_i. *Then* T *is zero on the union*

$$V = \bigcup_{i \in I} V_i.$$

Proof. Let $\phi \in C_c^\infty(V)$ and denote by K the support of ϕ. Since $(V_i)_{i \in I}$ is an open covering of the compact subset K, we can select a finite subcovering $V_{i_1}, \cdots, V_{i_r}$ of K. Let $(\psi_j)_{1 \le j \le r}$ be a C^∞ partition of unity subordinated to this covering (Theorem 1.1, Corollary 4). We have

$$\psi_j \in C_c^\infty(V_{i_j}),\ 0 \le \psi_j \le 1, \text{ and } \sum_{j=1}^{r} \psi_j = 1;$$

hence

$$\phi = \sum_{j=1}^{r} \phi \cdot \psi_j .$$

Since $\phi \cdot \psi_j \in C_c^\infty(V_{i_j})$ and T is zero on V_{i_j},

$$T(\phi) = \sum_{j=1}^{r} T(\phi \cdot \psi_j) = 0$$

which implies, ϕ being an arbitrary element of $C_c^\infty(V)$, that T is zero on V. Q.E.D.

Definition 2.6. *Let* $T \in \mathcal{D}'(\Omega)$. *The support of* T *is the complement in* Ω *of the largest open subset of* Ω *where* T *is zero. We denote it by* supp T.

Equivalently, a point belongs to the support of T if and only if there is no open neighborhood of it on which T is zero. Also, the support of T is the smallest closed subset of Ω outside of which the distribution T is zero.

As an exercise the reader can prove the following results:

1. If $\phi \in C_c^\infty(\Omega)$ and $T \in \mathcal{D}'(\Omega)$ are such that

$$\operatorname{supp} \phi \cap \operatorname{supp} T = \emptyset,$$

then

$$\langle T, \phi \rangle = 0.$$

2. Let $T \in \mathcal{D}'(\Omega)$. The support of T is the smallest closed subset F of U such that if $\phi, \psi \in C_c^\infty(\Omega)$ and $\phi = \psi$ on a neighborhood of F, then

$$\langle T, \phi \rangle = \langle T, \psi \rangle.$$

The next theorem establishes the complete relation between the two spaces of distributions $\mathcal{E}'(\Omega)$ and $\mathcal{D}'(\Omega)$.

Theorem 2.16. *Let* Ω *be an open set in* $\underline{R}^n$. *We have:* (1) $\mathcal{E}'(\Omega) \subset \mathcal{D}'(\Omega)$ *and the identity map is continuous with respect to the strong topologies;* (2) *the elements of* $\mathcal{E}'(\Omega)$ *are distributions with compact support in* Ω.

Proof. (i). We start by proving that $C_c^\infty(\Omega)$ is a dense subspace of $C^\infty(\Omega)$. Indeed, let (K_j) be an increasing sequence of compact subsets contained in Ω and such that their union is Ω and let (β_j) be a sequence of functions of $C_c^\infty(\Omega)$ such that $\beta_j = 1$ on a neighborhood of K_j. If $\phi \in C^\infty(\Omega)$, let $\phi_j = \beta_j \phi \in C_c^\infty(\Omega)$. It is easy to prove that

$$\phi_j \to \phi \textit{ in } C_c^\infty(\Omega).$$

(ii) Let $T \in \mathcal{E}'(\Omega)$. As we already know (Theorem 2.10), the identity map $C_c^\infty(\Omega) \to C^\infty(\Omega)$ is a continuous one, hence T defines a distribution on Ω.

On the other hand, in order to show that $\mathcal{E}'(\Omega)$ can be identified with a subspace of $\mathcal{D}'(\Omega)$ it suffices to show that if $T \in \mathcal{E}'(\Omega)$ is such that $\langle T, \phi \rangle = 0$ for all $\phi \in C_c^\infty(\Omega)$ then T must be the distribution identically zero. But this follows immediately from the fact that $C_c^\infty(\Omega)$ is dense in $C^\infty(\Omega)$.

To prove that the identity map from $\mathcal{E}'(\Omega)$ equipped with the strong topology into $\mathcal{D}'(\Omega)$ equipped with its strong topology is continuous, it suffices to observe, again, that the imbedding $C_c^\infty(\Omega) \to C^\infty(\Omega)$ is continuous; hence, every bounded set of $C_c^\infty(\Omega)$ is a bounded set of $C^\infty(\Omega)$.

(iii) Finally, let us show that every $T \in \mathcal{E}'(\Omega)$ has a compact support contained in Ω. In the proof of Theorem 2.14, we have seen that if $T \in \mathcal{E}'(\Omega)$, there is a number $\varepsilon > 0$, an integer $m \geq 0$ and a compact subset K of Ω such that for all $\phi \in C^\infty(\Omega)$ satisfying

$$p_{m,K}(\phi) \leq \varepsilon$$

we have

$$|\langle T, \phi\rangle| \leq 1.$$

We also remarked that $\langle T, \phi\rangle = 0$ for all $\phi \in C^\infty(\Omega)$ such that $p_{m,K}(\phi) = 0$. Since every $\psi \in C_c^\infty(\Omega - K)$ satisfies, trivially, this condition, it follows that T must be zero on $\Omega - K$; therefore the support of T must be contained in K. Q.E.D.

Examples. 1. The Dirac measure has compact support equal to $\{0\}$.

2. The elements of $C'(\Omega)$ are Radon measures (hence distributions) with compact support in Ω. (See Chapter 1, Problems 29 and 31.)

5. DERIVATIVES OF A DISTRIBUTION

Definition 2.7. Let Ω be an open subset of $\underline{R}^n$ and let T be an element of $\mathcal{D}'(\Omega)$. The partial derivative of T with respect to the variable x_k, $1 \leq k \leq n$, is the distribution $\partial T/\partial x_k$ defined by the formula

$$\left\langle\frac{\partial T}{\partial x_k}, \phi\right\rangle = -\left\langle T, \frac{\partial \phi}{\partial x_k}\right\rangle, \forall \phi \in C_c^\infty(\Omega). \tag{2.4}$$

This definition can be justified as follows. Let f be a function of class C^1 on Ω. Then f and all its partial derivatives $\partial_k f$ define distributions on Ω. An integration by parts shows that

$$\int_\Omega \frac{\partial f}{\partial x_k} \cdot \phi \, dx = - \int_\Omega f \cdot \frac{\partial \phi}{\partial x_k} \, dx, \; \forall \phi \in C_c^\infty(\Omega),$$

which coincides with (2.4) when T = f. Moreover, the last formula shows that the derivative of f in the *sense of distribution* [i.e., defined by (2.4)] coincides with the classical derivative of f whenever f is a differentiable function.

Let $\alpha = (\alpha_1, \cdots, \alpha_n)$ be an n-tuple of nonnegative integers and let T be an element of $\mathcal{D}'(\Omega)$. By induction we define $\partial^\alpha T$ as follows:

$$\langle \partial^\alpha T, \phi \rangle = (-1)^{|\alpha|} \langle T, \partial^\alpha \phi \rangle, \; \forall \phi \in C_c^\infty(\Omega). \tag{2.5}$$

If the distribution T is defined by a function f of class C^m on Ω then integrations by parts show that derivatives of f, in the sense of distributions, of order $|\alpha| \leq m$ coincide with the corresponding derivatives of f in the classical sense.

We list a few very useful properties of the derivatives of distributions:

1. *The differentiation, in the sense of distributions, is an everywhere defined operation on* $\mathcal{D}'(\Omega)$.
2. *Every distribution has derivatives of all orders.*
3. *For every* $T \in \mathcal{D}'(\Omega)$ *we have*

$$\frac{\partial^2 T}{\partial x_j \, \partial x_k} = \frac{\partial^2 T}{\partial x_k \, \partial x_j}, \quad 1 \leq j, k \leq n.$$

This result shows that we can interchange the order of derivatives of a distribution. As it is well known, this result is not true, in general, for the classical derivatives of functions.

4. *The map*

$$\partial_k : \mathcal{D}'(\Omega) \to \mathcal{D}'(\Omega), \ 1 \le k \le n,$$

is a continuous linear operator in the sense of the strong topologies.

Indeed, it is easy to see that the map

$$\partial_k : C_c^\infty(\Omega) \to C_c^\infty(\Omega)$$

is a continuous linear one; hence, it maps bounded sets into bounded sets. The continuity of the linear map ∂_k from $\mathcal{D}'(\Omega)$ into $\mathcal{D}'(\Omega)$ equipped with the strong topologies is a consequence of the following result, whose proof we leave to the reader: *If* A *is a bounded set of* $C_c^\infty(\Omega)$ *and if* B *is the image of* A *by* ∂_k, *then the image of the polar set* $B^\circ \subset \mathcal{D}'(\Omega)$ *by* ∂_k *is contained in* A°.

Examples. 1. As we remarked above, every function of $C^1(\Omega)$ defines a distribution on Ω whose partial derivatives of order one, in the classical sense, coincide with the corresponding partial derivatives in the sense of distributions.

2. The function $\ell og|x|$ is locally integrable in $\underline{R}^n$, hence it defines a distribution whose derivative is, by definition,

$$<\frac{d}{dx} \ell og|x|, \phi> = - \int_{-\infty}^{+\infty} (\ell og|x|)\phi'(x)\, dx, \ \forall \phi \in C_c^\infty(\underline{R}^n).$$

Hence $(d/dx)\ell og|x| = P{\cdot}V{\cdot}(1/x)$ (see Example 5 on p. 45).

3. Let $\Omega = \underline{R}$ and consider the Heaviside function

$$Y(x) = \begin{cases} 1 \text{ if } x > 0 \\ \\ 0 \text{ if } x < 0. \end{cases}$$

Its derivative in the sense of distributions is defined by

$$<\frac{dY}{dx}, \phi> = - <Y, \frac{d\phi}{dx}>$$

$$= - \int_0^\infty \frac{d\phi}{dx} = \phi(0) = <\delta, \phi>, \forall\phi \in C_c^\infty(\underline{R}).$$

Therefore,

$$Y' = \delta,$$

the Dirac measure.

Observe that the Heaviside function $Y(x)$ is discontinuous at the origin and the "jump" of discontinuity is equal to one. Thus, we can say that its derivative in the sense of distribution is the "jump" of discontinuity at the origin times the Dirac measure δ.

4. More generally, let Ω be an open interval (α, β) in $\underline{R}$, let $a \in \Omega$, and let f be a function of class C^1 on $\Omega - \{a\}$. Suppose that the following limits

$$\lim_{x \to a-} f(x) = f(a-0), \quad \lim_{x \to a+} f(x) = f(a+0)$$

exist and are finite. Denote by $Jf(a)$ the difference $f(a+0) - f(a-0)$. The function f has a *simple* discontinuity (or a discontinuity of the *first kind*) at the point a and $Jf(a)$ denotes the "jump" of f at the point a. Suppose also that $[f']$, the classical derivative of f, is a bounded function in $\Omega - \{a\}$.

From our assumptions, it follows that f and $[f']$ define distributions on Ω. Let us find the derivative of f in the sense of distributions. We have, for all $\phi \in C_c^\infty(\Omega)$,

$$<f', \phi> = -<f, \phi'> = - \int_\alpha^a f\cdot\phi' - \int_a^\beta f\cdot\phi'.$$

Integrating by parts we obtain

$$<f', \phi> = \left[f(a+0) - f(a-0)\right] \phi(a) + \int_\alpha^\beta [f']\cdot\phi$$

where [f'] denotes the classical derivative of f. Rewriting the last expression we get

$$<f', \phi> = <Jf(a)\delta_{(a)}, \phi> + <[f'], \phi>,$$

for all $\phi \in C_c^\infty(\Omega)$. Hence

$$f' = Jf(a)\delta_{(a)} + [f'],$$

i.e., the derivative of f in the sense of distributions is the sum of the classical derivative [f'] plus a *measure of mass* Jf(a) *concentrated at the point* a.

5. Let Ω be an open subset of $\underline{R}^2$ whose boundary Γ is a smooth curve. Let f(x, y) be a function of class C^1 on $\Omega \cup \Gamma$ which is zero on $(\Omega \cup \Gamma)^c$. Thus f and its partial derivatives have simple discontinuity on Γ. Denote by $\partial f/\partial x$ the derivative of f in the sense of distributions and by f'_x the classical derivative of f with respect to x. We have

$$<\frac{\partial f}{\partial x}, \phi> = -<f, \phi'_x> = -\iint_\Omega f(x, y)\phi'_x(x, y)\, dx\, dy.$$

by Green's formula we get

$$<\frac{\partial f}{\partial x}, \phi> = \iint_\Omega f'_x\cdot\phi\, dx\, dy + \int_\Omega f(x, y)\phi(x, y) \cos\alpha\, d\ell$$

where α is the angle between the outer normal to Γ and the x axis.

If we define

$$<\mu, \phi> = \int_\Gamma f(x, y)\phi(x, y) \cos\alpha\, d\ell,$$

we can interpret μ as a *measure on* $\underline{R}^2$ *supported by* Γ *and given by the density* $f(x, y) \cos \alpha$. Replacing above we get

$$<\frac{\partial f}{\partial x}, \phi> = <f'_x, \phi> + <\mu, \phi>$$

Therefore, we can say that the derivative of f in the sense of distributions is equal to the classical derivative plus a measure concentrated at the boundary Γ. A similar example can be given in a three-dimensional (n-dimensional) space.

6. Going back to the Example 3), we can say that the Heaviside function is a solution of the differential equation

$$\frac{dY}{dx} = \delta.$$

Let us define the Heaviside function in $\underline{R}^n$ as follows:

$$Y(x_1, \cdots, x_n) = \begin{cases} 1 \text{ if } x_1 \geq 0, \cdots, x_n \geq 0 \\ \\ 0 \text{ elsewhere.} \end{cases}$$

Then, $Y(x_1, \cdots, x_n)$ is a solution of the following partial differential equation:

$$\frac{\partial^n Y}{\partial_{x_1} \cdots \partial_{x_n}} = \delta.$$

Let us check this in the case $n = 2$. By definition,

$$<\frac{\partial^2 Y}{\partial_{x_1} \partial_{x_2}}, \phi> = <Y, \frac{\partial^2 \phi}{\partial_{x_1} \partial_{x_2}}>$$

for all $\phi \in C_c^\infty(\underline{R}^2)$. On the other hand,

$$<Y, \frac{\partial^2 \phi}{\partial_{x_1} \partial_{x_2}}> = \int_0^\infty \int_0^\infty \frac{\partial^2 \phi}{\partial_{x_1} \partial_{x_2}} = \phi(0, 0) = <\delta, \phi>;$$

hence, $\partial^2 Y/\partial x_1 \partial x_2 = \delta$. Q.E.D.

Definition 2.8. *Let* $P = P(\partial) = \sum_{|p|\le m} a_p \partial^p$ *be a partial differential operator with constant coefficients in* $\underline{R}^n$. *We say that a distribution* $E \in \mathcal{D}'(\underline{R}^n)$ *is a fundamental (or elementary) solution of the partial differential operator* P *if* $P(\partial)E = \delta$.

If

$$ {}^tP = {}^tP(\partial) = \sum_{|p|\le m} (-1)^{|p|} a_p \partial^p $$

is the *transpose* of the operator P, then $E \in \mathcal{D}'(\underline{R}^n)$ is a fundamental solution of P if and only if

$$ \langle E, {}^tP(\partial)\phi\rangle = \phi(0), \quad \phi \in C_c^\infty(\underline{R}^n). $$

The Heaviside functions defined in Examples 3 and 6 above are, respectively, fundamental solutions of the operators d/dx and $\partial^n/\partial x_1 \cdots \partial x_n$. In Chapter 7, we shall give more examples of fundamental solutions and we shall prove Malgrange's theorem on the existence of fundamental solutions of partial differential operators with constant coefficients.

6. NORMAL SPACES OF DISTRIBUTION

Definition 2.9. *We say that a topological vector space* E *is a normal space of distributions on* $\Omega \subset \underline{R}^n$ *if*: (i) $C_c^\infty(\Omega) \subset E \subset \mathcal{D}'(\Omega)$ *with continuous imbeddings, where we suppose* $\mathcal{D}'(\Omega)$ *equipped with its strong topology;* (ii) $C_c^\infty(\Omega)$ *is dense in* E.

If E is a normal space of distributions, condition (ii) implies that its dual E' can be identified with a subspace of $\mathcal{D}'(\Omega)$. Condition (i) implies that $E'_b \subset \mathcal{D}'(\Omega)$ with continuous imbedding.

Examples. 1. $C_c^\infty(\Omega)$ is trivially a normal space of distributions on Ω.

2. In the proof of Theorem 2.16, we have seen that $C_c^\infty(\Omega) \subset C^\infty(\Omega) \subset \mathcal{D}'(\Omega)$ with continuous imbeddings and that $C_c^\infty(\Omega)$ is dense in $C^\infty(\Omega)$; hence $C^\infty(\Omega)$ is a normal space of distribution and $\mathcal{E}'(\Omega)$ is a subspace of $\mathcal{D}'(\Omega)$.

3. The spaces $L^p(\Omega)$, $1 \le p < +\infty$, are normal spaces of distributions. The fact that $C_c^\infty(\Omega)$ is dense in $L^p(\Omega)$ is a consequence of Theorem 1.1. Since $L^\infty(\Omega)$ is the dual of $L^1(\Omega)$, it follows that $L^\infty(\Omega)$ is a subspace of $\mathcal{D}'(\Omega)$.

4. The space $C_c^m(\Omega)$, $0 \le m \le +\infty$, is a normal space of distributions. The fact that $C_c^\infty(\Omega)$ is dense in $C_c^m(\Omega)$ is again a consequence of Theorem 1.1.

7. THE SPACE OF DISTRIBUTIONS OF FINITE ORDER

Let us consider the space $C_c^m(\Omega)$ equipped with its natural topology (Section 2, complements) and denote by $\mathcal{D}'^m(\Omega)$ its dual. Since $C_c^m(\Omega)$ is a normal space of distributions, $\mathcal{D}'^m(\Omega)$ is a subspace of $\mathcal{D}'(\Omega)$ for all $m \ge 0$. On the other hand, $C_c^{m+1}(\Omega) \subset C_c^m(\Omega)$ with continuous imbedding and $C_c^{m+1}(\Omega)$ is dense in $C_c^m(\Omega)$; hence $\mathcal{D}'^m(\Omega) \subset \mathcal{D}'^{m+1}(\Omega)$, with continuous imbeddings in the strong topologies.

Definition 2.10. *We say that a distribution* $T \in \mathcal{D}'(\Omega)$ *has order* $\le m$ *if* $T \in \mathcal{D}'^m(\Omega)$. *We say that* T *has order* m *if* m *is the smallest integer such that* $T \in \mathcal{D}'^m(\Omega)$.

Examples. 1. The Dirac measure is a distribution of order zero. Distributions of order zero are Radon measures.

2. If $|p| = m$, $\partial^p\delta = \delta^{(p)}$ is a distribution of order m.

3. Every distribution $T \in \mathcal{E}'(\Omega)$ is of *finite order*. This fact is a consequence of Theorem 2.   to be proved in Section 9.

Let $\mathcal{D}_F(\Omega)$ be the space $C_c^\infty(\Omega)$ equipped with the coarsest topology for which the imbedding $C_c^\infty(\Omega) \to C_c^m(\Omega)$ is continuous for all $m \ge 0$. We claim that $\mathcal{D}_F(\Omega)$ is a *normal space of distributions*. First of all, the identity map from $C_c^\infty(\Omega)$ into $\mathcal{D}_F(\Omega)$ is continuous. Indeed, it suffices to show that for every compact set $K \subset \Omega$ the identity map $C_c^\infty(\Omega; K) \to \mathcal{D}_F(\Omega)$ is continuous (Proposition 1.1). Since the topology of $\mathcal{D}_F(\Omega)$ is the coarsest one for which the imbedding $C_c^\infty(\Omega) \to C_c^m(\Omega)$ is continuous for all $m \geq 0$, it suffices to prove, by a known result of general topology, that the identity map $C_c^\infty(\Omega; K) \to C_c^m(\Omega)$ is continuous for all $m \geq 0$, which is trivial. Obviously, $C_c^\infty(\Omega)$ is dense in $\mathcal{D}_F(\Omega)$. Finally, the imbedding $\mathcal{D}_F(\Omega) \to \mathcal{D}'(\Omega)$ decomposes into the imbedding $\mathcal{D}_F(\Omega) = C_c^\infty(\Omega) \to C_c^m(\Omega)$, which is continuous by definition, and the imbedding $C_c^m(\Omega) \to \mathcal{D}'(\Omega)$, which is also continuous, because $C_c^m(\Omega)$ is a normal space of distributions.

If we denote by $\mathcal{D}_F'(\Omega)$ the dual of $\mathcal{D}_F(\Omega)$, then $\mathcal{D}_F'(\Omega)$ is a subspace of $\mathcal{D}'(\Omega)$ and furthermore

$$\mathcal{D}_F'(\Omega) = \bigcup_{m \geq 0} \mathcal{D}'^m(\Omega),$$

i.e., $\mathcal{D}_F'(\Omega)$ is the *space of distributions of finite order.*

In general, the spaces $\mathcal{D}_F'(\Omega)$ and $\mathcal{D}'(\Omega)$ are distinct. In fact, let $\Omega = \underline{R}$, let δ_k be the Dirac measure at the point $k \in \underline{N}$ [i.e., $\delta_k(\phi) = \phi(k)$ for all $\phi \in C_c^\infty(\underline{R})$], and let $\delta_k^{(k)}$ be the kth derivative of δ_k. It is easily seen that the series $T = \sum_{k \geq 0} \delta_k^{(k)}$ defines a distribution on $\underline{R}$ and $T \notin \mathcal{D}_F'(R)$. This example also shows that the topology of $\mathcal{D}_F(R)$ is *strictly coarser* than the natural topology of $C_c^\infty(\underline{R})$.

8. ON SOME PROPERTIES OF DISTRIBUTIONS DEFINED ON THE REAL LINE

As an application of the definition of derivative of a distribution, we can define the notion of *primitive* or *indefinite integral* of a distribution. The theory is particularly simple in the case of one variable. We shall discuss it here, leaving aside the case of more variables. Both cases are discussed in Schwartz's book [28, Chapter II].

Definition 2.11. *We say that* $T \in \mathcal{D}'(\underline{R})$ *is a primitive of* S *if* $\mathcal{D}'(R)$ *if* $dT/dx = S$.

Theorem 2.17 *Every distribution on* $\underline{R}$ *has infinitely many primitives, two of which differ by a constant.*

Proof. 1. Denote by H the subspace of all functions $\chi \in C_c^\infty(\underline{R})$ such that

$$\int_{-\infty}^{+\infty} \chi(t)\, dt = 0.$$

A function χ belongs to H if and only if there is $\psi \in C_c^\infty(\underline{R})$ such that $\chi = \psi'$. Indeed, if $\chi \in H$, the function

$$\psi(x) = \int_{-\infty}^{x} \chi(t)\, dt$$

belongs, obviously, to $C_c^\infty(\underline{R})$. Conversely, if $\chi = \psi'$, with $\psi \in C_c^\infty(\underline{R})$, then χ must satisfy the above condition.

2. The subspace H is a hyperplane (i.e., its codimension is one) of $C_c^\infty(\underline{R})$ because it is defined as the set of all solutions of a homogeneous linear equation. Choose $\phi_0 \in C_c^\infty(\underline{R})$ such that

$$\int_{-\infty}^{+\infty} \phi_0(t)\, dt = 1.$$

Then, every $\phi \in C_c^\infty(\underline{R})$ can be written in a unique way as

$$\phi = \lambda\phi_0 + \chi$$

with

$$\lambda = \int_{-\infty}^{+\infty} \phi(t)\, dt$$

and $\chi \in H$, i.e., $\chi = \psi'$ for some $\psi \in C_c^\infty(\underline{R})$.

3. Next, it is very easy to check that a distribution T is a primitive of a given $S \in \mathcal{D}'(\underline{R})$ if and only if, for every $\chi \in H$, we have

$$T(\chi) = -S(\psi)$$

where $\chi = \psi'$.

4. If $S \in \mathcal{D}'(\underline{R})$ is given, define

$$T(\phi) = \lambda T(\phi_0) - S(\psi), \ \forall \ \phi \in C_c^\infty(\underline{R})$$

where $T(\phi_0)$ is an arbitrary constant. It is clear from this definition that $T(\chi) = -S(\psi)$ for every $\chi \in H$. It then suffices to prove that T is a distribution on $\underline{R}$. According to Theorem 2.13 it suffices to show that if (ϕ_k) is a sequence converging to zero in $C_c^\infty(\underline{R})$, the numerical sequence $(T(\phi_k))$ converges to zero. If the sequence (ϕ_k) converges to zero in $C_c^\infty(\underline{R})$, all the functions ϕ_k have their support contained in a fixed compact subset of $\underline{R}$ and they converge, together with their partial derivatives, uniformly to zero in $\underline{R}$. Hence,

$$\lambda_k = \int_{-\infty}^{+\infty} \phi_k(t)\ dt \to 0 \text{ as } k \to +\infty.$$

As a consequence, we have

$$\chi_k = \phi_k - \lambda_k\phi_0 \to 0 \text{ in } C_c^\infty(\underline{R}) \text{ as } k \to +\infty.$$

Therefore,

$$T(\phi_k) = \lambda_k T(\phi_0) - S(\psi_k) \to 0 \text{ as } k \to +\infty.$$

5. Finally, suppose that T_1 and T_2 are two primitives of S. Then, we have

$$\langle T_1 - T_2, \phi\rangle = \lambda\langle T_1 - T_2, \phi_0\rangle.$$

Letting $C = \langle T_1 - T_2, \phi_0 \rangle$ and recalling the definition of λ, we have

$$\langle T_1 - T_2, \phi \rangle = C \cdot \int_{-\infty}^{+\infty} \phi(t)\, dt = \langle C, \phi \rangle, \forall \phi \in C_c^\infty(\underline{R});$$

therefore,

$$T_1 - T_2 = C. \quad \text{Q.E.D.}$$

As a consequence of Theorem 2.17, it is obvious that a distribution on $\underline{R}$ which has a derivative equal to zero is a constant. Furthermore, if $T \in \mathcal{D}'(\underline{R})$ is such that

$$\frac{d^2T}{dx^2} = 0,$$

then T is a polynomial Ax + B. Indeed, set S = dT/dx. Since

$$\frac{dS}{dx} = 0,$$

S is equal to a constant A. Now comparing the two distributions T and Ax, we see that they have the same derivative S, hence they differ by a constant B. Thus T = Ax + B. Q.E.D.

If we call the distribution T the *primitive of order* p of the distribution S, where

$$\frac{d^pT}{dx^p} = S,$$

we can summarize the above results in the following.

Corollary. Every distribution on the real line has infinitely many primitives of order p. *Two of such primitives differ by a polynomial of degree* $\leq$ p-1.

Definition 2.12. *A function* f *defined on* [a, b] *is said to be absolutely continuous if, given* $\varepsilon > 0$, *there is* $\delta > 0$ *such that*

$$\sum_{i=1}^{n} |f(x_i) - f(y_i)| < \varepsilon$$

for every finite collection of nonoverlapping intervals (x_i, y_i) *with*

$$\sum_{i=1}^{n} |x_i - y_i| < \delta$$

Every absolutely continuous function is continuous. A function is absolutely continuous if and only if it is an *indefinite integral* in the sense of Lebesgue's theory of integration [25, p. 106].

Theorem 2.18. *Let* $f(x)$ *be a continuous function defined on* $\underline{R}$ *and suppose that*

$$f'(x) = g(x) \quad \text{a.e.}$$

and that $g(x)$ *is locally integrable. Then the derivative of* f *in the sense of distributions is* g.

Proof. We have

$$\langle f', \phi \rangle = -\langle f, \phi' \rangle = -\int_{-\infty}^{+\infty} f(x)\phi'(x)\,dx$$

for all $\phi \in C_c^\infty(\underline{R})$. Integrating by parts, we get

$$\langle f', \phi \rangle = -f\cdot\phi\Big|_{-\infty}^{+\infty} + \int_{-\infty}^{+\infty} f'(x)\phi(x)\,dx$$

$$= \int_{-\infty}^{+\infty} g(x)\phi(x)\,dx = \langle g, \phi \rangle;$$

hence $f' = g$ as distributions. Q.E.D.

Theorem 2.19. *If the derivative of* $T \in \mathcal{D}'(\underline{R})$ *is a locally integrable function* $g(x)$ *then* T *is an absolutely continuous function.*

Proof. Since g is, by assumption, a locally integrable function, the integral

$$f(x) = \int_a^x g(t)\, dt$$

is well defined and f is a *primitive* of g, i.e.,

$$f'(x) = g(x) \text{ a.e.},$$

[25, p. 103]. Furthermore, f is an absolutely continuous function [25, p. 106].

By Theorem 2.18, the derivative of f is the sense of distributions is also g, hence the distribution T - f equals a constant, which implies that T is an absolutely continuous function. Q.E.D.

These two theorems can be extended to the case of functions of several variables [28, Chapter II, Theorem V].

9. THE LOCAL STRUCTURE OF DISTRIBUTIONS

In Section 5 we have introduced the notion of derivative of a distribution and we have seen that every distribution has derivatives of all orders. The new notion of derivative allows us to differentiate, in particular, locally integrable functions and their derivatives as distributions. In the next theorem we shall prove that *locally* every distribution is the derivative (in the sense of distributions) of a bounded function. We shall follow Schwartz [28, Chapter III, Theorem XXI].

Theorem 2.20. *Let* $T \in \mathcal{D}'(\Omega)$ *and let* ω *be a relatively compact open set such that* $\bar{\omega} \subset \Omega$. *Then, we can find a function* $f \in L^\infty(\omega)$ *and an integer* $m \geq 0$ *such that*

$$T = \frac{\partial^{mn} f}{\partial x_1^m \cdots \partial x_n^m} \quad on \quad \omega.$$

Proof. 1. Let $K = \bar{\omega}$. Since T is a distribution, T is a continuous linear functional on $C_c^\infty(\Omega; K)$. Hence, given $\varepsilon > 0$, we can find a neighborhood

$$V = V(K, k, \eta) = \{\phi \in C_c^\infty(\Omega; K) : |\partial^\alpha \phi(x)| \le \eta, \forall\, |\alpha| \le k, \forall\, x \in K\} \tag{2.5}$$

such that

$$|T(\phi)| \le \varepsilon, \forall \phi \in V. \tag{2.6}$$

2. In order to simplify our notations, let us denote, with m a nonnegative integer, by $\partial^m/\partial x^m$ the partial derivative

$$\frac{\partial^{mn}}{\partial x_1^m \cdots \partial x_n^n} .$$

Let E be the subspace of all functions of the form

$$\psi = \frac{\partial^{k+1} \phi}{\partial x^{k+1}}, \ \forall \phi \in C_c^\infty(\Omega; K). \tag{2.7}$$

Since the functions have compact support, the correspondence between ϕ and ψ given by (2.7) is one-to-one.

Let $L^1(K)$ be the Banach space of integrable functions on K and let us consider on E the topology induced by $L^1(K)$. We claim that *if a sequence of functions*

$$\psi_j = \frac{\partial^{k+1} \phi_j}{\partial x^{k+1}}$$

converges to zero in $L^1(K)$ *then the sequence* $T(\phi_j)$ *converges to zero.* Indeed, let Q be a hypercube with length side $\ell \geq 1$ containing K. For any $\eta > 0$, there is an index j_0 such that

$$\int \cdots \int_K \left| \frac{\partial^{k+1} \phi_j}{\partial x^{k+1}} \right| dx_1 \cdots dx_n \leq \frac{\eta}{\ell^{(k+1)n}}, \tag{2.8}$$

for all $j \geq j_0$. Writing

$$\frac{\partial^k \phi_j}{\partial x^k}(x) = \int_{-\infty}^{x_1} \cdots \int_{-\infty}^{x_n} \frac{\partial^{k+1} \phi_j}{\partial t^{k+1}} dt_1 \cdots dt_n \tag{2.9}$$

then (2.8) easily implies

$$\left| \frac{\partial^k \phi_j}{\partial x^k}(x) \right| \leq \frac{\eta}{\ell^{kn}}, \tag{2.10}$$

since $\ell \geq 1$. Now, using the integral representation (2.9) several times, it is easy to see that the inequality (2.10) implies the following one:

$$|\partial^\alpha \phi_j(x)| \leq \eta, \ \forall\, x \in K, \forall\, |\alpha| \leq k, \forall\, j \geq j_0, \tag{2.11}$$

which shows that $\phi_j \in V$, for all $j \geq j_0$ and, consequently, $|T(\phi_j)| \leq \varepsilon$, $\forall\, j \geq j_0$. Hence, our contention follows immediately.

3. Define on E the linear functional

$$L(\psi) = T(\phi)$$

where ϕ and ψ are related by (2.7). By what we have just proved, L is a continuous linear functional on E equipped with the topology induced by $L^1(K)$. Since E is obviously a proper linear subspace of $L^1(K)$, the linear functional L can be extended, according to the Hahn-Banach theorem, to a continuous linear functional $\tilde{L}$ on $L^1(K)$. By a well-known characterization of continuous linear functionals on $L^1(K)$ [25, p. 246] there is an element $g \in L^\infty(K)$ such that

$$\tilde{L}(\psi) = \int \cdots \int_K g \cdot \psi, \forall\, \psi \in E.$$

Consequently, for every $\phi \in C_c^\infty(\Omega; K)$, *a fortiori* for every $\phi \in C_c^\infty(\omega)$, we have

$$T(\phi) = L(\psi) = \int \cdots \int_K g \cdot \frac{\partial^{k+1}\phi}{\partial x^{k+1}} = \langle (-1)^{(k+1)n} \frac{\partial^{k+1} g}{\partial x^{k+1}}, \phi \rangle,$$

which implies, setting $m = k+1$ and $f = (-1)^{mn} g$, that

$$T = \frac{\partial^m f}{\partial x^m} \text{ on } \omega \text{ Q.E.D.}$$

Remarks. 1. The function f is not necessarily unique. Indeed, one can add to f any solution of

$$\frac{\partial^m h}{\partial x^m} = 0.$$

2. Set $f = 0$ on K^c and define

$$F(x_1, \cdots, x_n) = \int_{-\infty}^{x_1} \cdots \int_{-\infty}^{x_n} f(t_1, \cdots, t_n)\, dt_1 \cdots dt_n.$$

Clearly, F is a continuous function and $\partial^n F/\partial x_1 \cdots \partial x_n = f$. Making the appropriate replacements above, we get

$$T = \frac{\partial^{m+1} F}{\partial x^{m+1}} \text{ on } \omega.$$

Therefore, the distribution T can be represented, locally, as *a derivative of a continuous function.*

3. Let U be an open neighborhood of $K = \bar{\omega}$ such that $U \subset \Omega$.

Let $\alpha \in C_c^\infty(\Omega)$ with $\alpha = 1$ on K and let $G = \alpha F$. It is clear that

$$T = \frac{\partial^{m+1} G}{\partial x^{m+1}} \quad \text{on } \omega ,$$

where G is a *continuous function with compact support contained in an open neighborhood* U *of* K.

We can summarize these results in the following result.

Theorem 2.21. *Let* $T \in \mathcal{D}'(\Omega)$ *and let* ω *be a relatively compact open set such that* $\bar{\omega} \subset \Omega$. *The distribution* T *coincides, on* ω, *with the derivative of a continuous function having its support in an arbitrary neighborhood of* $\bar{\omega}$.

Remark. Theorems 2.20 and 2.21 are of a local nature. If Ω is an open set *not* relatively compact, it is not true, in general, that a distribution on Ω be a derivative of a function. Indeed, the distribution $T = \sum_{k\geq 0} \delta_k^{(k)}$ on $\underline{R}$ defined at the end of Section 7 is a sum of derivatives of measures.

For distributions with compact support, we can get a *global representation in* Ω.

Theorem 2.22. *Every distribution* $T \in \mathcal{E}'(\Omega)$ *can be represented, in many ways, as follows:*

$$T = \sum_{|p|\leq r} \partial^p f_p \quad \text{on } \Omega,$$

where f_p *are continuous functions with compact support contained in an arbitrary neighborhood* U *of* K, *the support of* T.

Proof. Let ω be a relatively compact open set such that

$$K \subset \omega \subset \bar{\omega} \subset U \subset \Omega .$$

By Theorem 2.21, there is a continuous function f with compact support in U such that

$$T = \partial^r f \text{ on } \omega,$$

i.e.,

$$\langle T, \psi\rangle = (-1)^{|r|} \int \cdots \int_{\Omega} f \cdot \partial^r \psi, \forall \psi \in C_c^{\infty}(\omega).$$

Let $\alpha \in C_c^{\infty}(\omega)$ be equal to one on a neighborhood of K. Then, for every $\phi \in C^{\infty}(\Omega)$ we have

$$\langle T, \phi\rangle = \langle T, \alpha\phi\rangle = (-1)^{|r|} \int \cdots \int_{\Omega} f \cdot \partial^r (\alpha\phi).$$

By the Leibniz formula (see Chapter 1, Problem 3)

$$\partial^r(\alpha \cdot \phi) = \sum_{p \leq r} \frac{r!}{p!(r-p)!} \partial^{r-p}\alpha \cdot \partial^p \phi.$$

Hence,

$$\langle T, \phi\rangle = (-1)^{|r|} \sum_{p \leq r} \frac{r!}{p!(r-p)!} \int \cdots \int_{\Omega} f \cdot \partial^{r-p}\alpha \cdot \partial^p \phi.$$

Letting

$$f_p = (-1)^{|r| + |p|} \frac{r!}{p!(r-p)!} f \cdot \partial^{r-p}\alpha$$

and integrating by parts we get

$$\langle T, \phi\rangle = \int \cdots \int_{\Omega} \left(\sum_{p \leq r} \partial^p f_p\right) \phi$$

for all $\phi \in C^{\infty}(\Omega)$. Q.E.D.

Corollary. Every distribution with compact support is of finite order.

Proof. If $T = \partial^p f$, where f is a locally integrable function in $\underline{R}^n$, T defines a continuous linear functional on $C_c^{|p|}(\underline{R}^n)$; hence T is of finite order. Q.E.D.

Examples. 1. We have (see Section 5, Example 6) that the Dirac measure δ can be represented as follows:

$$\delta = \frac{\partial^n Y}{\partial x_1 \cdots \partial x_n}$$

where Y, the Heaviside function, is a bounded function in $\underline{R}^n$.

2. Let K be a compact set in $\underline{R}$. The characteristic function of K, $\chi_K(t)$, defines an element of $\mathcal{E}'(\underline{R})$. The function

$$f(x) = \int_{-\infty}^{x} \chi_K(t)\, dt$$

is absolutely continuous and by Theorem 2.14 we have

$$\frac{df}{dx} = \chi_K,$$

i.e., the distribution χ_K can be represented as a derivative of a continuous function.

From the definition of support of a distribution it immediately follows that if $\phi \in C_c^\infty(\Omega)$ is zero on a neighborhood of the support of $T \in \mathcal{D}'(\Omega)$ then $\langle T, \phi\rangle = 0$. We are going to see that this result can be considerably improved when T has compact support.

Theorem 2.23. Let T *be a distribution with compact support* $K \subset \Omega$ *and suppose that* T *has order* $\leq m$. *Then,* $\langle T, \phi\rangle = 0$, *for all* $\phi \in C^\infty(\Omega)$ *such that* $\partial^p\phi = 0$ *on* K, *for all* $|p| \leq m$.

Proof. 1. Let K_ε be the ε-neighborhood of K. Since, by assumption, all derivatives $\partial^p\phi$ with $|p| = m$ vanish on K, given $\eta > 0$ we can find $\varepsilon > 0$ sufficiently small so that $K_\varepsilon \subset \Omega$ and

$$|\partial^p\phi(x)| \leq \eta, \ \forall \ x \in K_\varepsilon.$$

For every $x \in K$, let $x_0 \in K$ be such that the distance between x_0 and x is at most ε. Then, if $|p| = m - 1$, we can write [since $\partial^p\phi(x_0) = 0$]

$$\partial^p\phi(x) = \int_0^1 \sum_{j=1}^n \partial_j\partial^p\phi(x_0 + t(x-x_0))\cdot(x_j-x_{0,j})\ dt;$$

hence we get

$$|\partial^p\phi(x)| \leq (\varepsilon\sqrt{n})\eta.$$

Proceeding by induction, we obtain

$$|\partial^p\phi(x)| \leq (\varepsilon\sqrt{n})^{m-|p|}\cdot\eta, \ \forall\ |p| \leq m. \tag{2.12}$$

2. With a proof similar to that of Corollary 3 of Theorem 1.1, we can find a function α_ε with the following properties: $\alpha_\varepsilon \in C_c^\infty(\Omega)$, supp $\alpha_\varepsilon \subset K_\varepsilon$, $\alpha_\varepsilon = 1$ on $K_{\varepsilon/4}$, and

$$|\partial^p\alpha_\varepsilon(x)| \leq C(p, n)\varepsilon^{-|p|}, \ \forall x.$$

By choosing $C > 0$ sufficiently large, we get

$$|\partial^p\alpha_\varepsilon(x)| \leq C\cdot\varepsilon^{-|p|}, \forall\ x, \forall\ |p| \leq m. \tag{2.13}$$

3. Set $\psi_\varepsilon = \alpha_\varepsilon\phi$. By Leibniz's formula we have

$$\partial^p \psi_\varepsilon = \sum_{q \le p} \frac{p!}{q!(p-q)!} \partial^{p-q} \alpha_\varepsilon \cdot \partial^q \phi.$$

Taking into account the inequalities (2.12) and (2.13), we obtain the estimate

$$|\partial^p \psi_\varepsilon(x)| \le C\varepsilon^{-|p|+|q|} \cdot (\varepsilon\sqrt{n})^{m-|q|} \cdot \eta \le C \cdot \eta \tag{2.14}$$

for all x and for all $|p| \le m$, where C is a suitable constant depending on m and n, but not on p.

4. Since $\psi_\varepsilon = \phi$ on a neighborhood of K the support of T, we have

$$\langle T, \psi_\varepsilon \rangle = \langle T, \phi \rangle. \tag{2.15}$$

On the other hand, since $T \in \mathcal{D}'^m(\Omega)$, by Theorem 2.12, there is a constant $C_1 > 0$ such that

$$|\langle T, \chi \rangle| \le C_1 \sup_{\substack{x \in \Omega \\ |p| \le m}} |\partial^p \chi|, \forall \chi \in C_c^\infty(\Omega, K_\varepsilon)$$

Replacing χ by ψ_ε and taking into account (2.14) and (2.15), we get

$$|\langle T, \phi \rangle| \le C_1 C \eta$$

which implies, η being an arbitrary positive number, that $\langle T, \phi \rangle = 0$. Q.E.D.

As a consequence we can characterize the distributions whose support is the origin, namely by the following theorem.

Theorem 2.24. *Every distribution* T *whose support is the origin can be represented in a unique way as a finite linear combination of derivatives of the Dirac measure.*

Proof. Since T has compact support, by the corollary of Theorem 2.22, T is a distribution of finite order, i.e., $T \in \mathcal{D}'^m$ for some $m \in \underline{N}$.

On the other hand, every $\phi \in C^\infty$ can be written as

$$\phi(x) = \sum_{|p|\le m} \frac{\partial^p \phi(0)}{p!} x^p + R_m(x)$$

where R_m is a C^∞ function such that $\partial^p R_m(0) = 0$ for all $|p| \le m$. By Theorem 2.23, it follows that

$$\langle T, \phi\rangle = \sum_{|p|\le m} \frac{\partial^p \phi(0)}{p!} \langle T, x^p\rangle.$$

By setting

$$c_p = \frac{(-1)^{|p|}\langle T, x^p\rangle}{p!}$$

we get

$$T = \sum_{|p|\le m} c_p \partial^p \delta.$$

The fact that the decomposition is unique follows by noticing that if T = 0, then $\langle T, x^p\rangle = 0$; hence $c_p = 0, \forall p$. Q.E.D.

PROBLEMS

1. Prove that

$$p_{m,j}(\phi) = \sup_{\substack{x\in K_j \\ |\alpha|\le m}} |\partial^\alpha \phi(x)|, \quad m \ge 0,\ j = 1, 2, \cdots,$$

define a seminorm on $C^\infty(\Omega)$, where (K_j) is an increasing sequence of compact subsets of Ω whose union is Ω.

2. Prove that the natural topology on $C^\infty(\Omega)$ does not change if we replace the sequence (K_j) by another increasing sequence of compact subsets of Ω whose union is Ω.

3. Show that the family of sets

$$V = V(m, j, \varepsilon) = \{\phi \in C^\infty(\Omega): p_{m,j}(\phi) < \varepsilon\}$$

with $m \geq 0$, $j = 1, 2, \cdots$, and $\varepsilon > 0$ form a fundamental system of convex balanced absorbing open neighborhoods of zero in $C^\infty(\Omega)$.

4. Prove the equivalence of the following conditions: (i) (ϕ_k) is a sequence converging to zero in $C^m(\Omega)$; (ii) for every $0 \leq |\alpha| \leq m$ the sequence $(\partial^\alpha \phi_k)$ converges to zero in $C(\Omega)$; (iii) for every $0 \leq |\alpha| \leq m$ the sequence $(\partial^\alpha \phi_k)$ converges to zero on every compact subset of Ω.

5. Prove that the natural topology of $C^m(\Omega)$ is the coarsest one for which the linear map

$$\partial^\alpha : C^m(\Omega) \to C(\Omega)$$

is continuous, $\forall |\alpha| \leq m$, where $C(\Omega)$ is equipped with its natural topology.

6. Prove that: (i) if $u \in C^m(\underline{R}^n)$ then $\alpha_\varepsilon * u \in C^\infty(\underline{R}^n)$ converges to u in C^m as $\varepsilon \to 0$, where α_ε is the test function defined in Chapter 1, Section 2; (ii) if $u \in C_c^m(\underline{R}^n)$ then $\alpha_\varepsilon * u \in C_c^\infty(\underline{R}^n)$ converges to u in C_c^m.

7. Prove that the natural topology on $C^\infty(\Omega)$ is the coarsest one for which the identity map $C^\infty(\Omega) \to C^m(\Omega)$ is continuous, $\forall m \geq 0$.

8. A subset A of $C^\infty(\Omega)$ is bounded if and only if, for every α, the set $\partial^\alpha A = \{\partial^\alpha \phi: \phi \in A\}$ is bounded in $C(\Omega)$.

9. Let Ω be an open subset of $\underline{C}$ and denote by $H(\Omega)$ the vector space of all holomorphic functions on Ω. Prove that: (i) on $H(\Omega)$ the topologies induced by $C(\Omega)$ and by $C^\infty(\Omega)$ do coincide (*hint:* use Cauchy's formula); (ii) $H(\Omega)$ is a Montel space.

10. Prove that: (i) $p_m(\phi) = \sup_{x \in K, |\alpha| \leq m} |\partial^\alpha \phi(x)|$, $m \in \underline{N}$, is a norm on $C_c^\infty(\Omega; K)$; (ii) the topology defined on $C_c^\infty(\Omega; K)$ by the sequence of norms $(p_m)_{m \in \underline{N}}$ coincides with the one induced by $C^\infty(\Omega)$.

11. Prove that the identity map $C_c^\infty(\Omega) \to C_c^m(\Omega)$ is continuous, $\forall$ $m \in \underline{N}$, and that $C_c^\infty(\Omega)$ is dense in $C_c^m(\Omega)$.

12. Prove that $C_c^\infty(\Omega; K)$ is a closed subspace of $C^\infty(\Omega)$.

13. Prove Theorem 2.8.

14. If $\alpha \in C^\infty(\Omega)$, show that the linear map $C_c^\infty(\Omega) \ni \phi \to \alpha \cdot \phi \in C_c^\infty(\Omega)$ is continuous.

15. Prove that the derivative ∂_k, $1 \leq k \leq n$, is a continuous linear map from $C_c^\infty(\Omega)$ into $C_c^\infty(\Omega)$.

16. Prove that the spaces $L^p_{\ell oc}(\Omega)$ and $L^p_c(\Omega)$ are normal spaces of distributions.

17. Is $L^\infty(\Omega)$ a normal space of distributions?

18. If Ω is an open set in $\underline{R}^n$, prove that $C^m(\Omega)$ is a normal space of distributions. Its dual $E'^m(\Omega)$ is the space of distributions of order m with compact support in Ω.

19. Let x be a real variable. Compute, in the sense of distributions, the successive derivatives of $|x|$.

20. Let Y(x) be the Heaviside function on $\underline{R}$. Prove that: (i) $Y(x)e^{x}$ is a fundamental solution of the operator $P(d/dx) = (d/dx) - \lambda$, λ a complex number; (ii) $Y(x)e^{\lambda x} x^{m-1}/(m-1)!$ is a fundamental solution of $[(d/dx) - \lambda]^m$; (iii) $Y(x)\ (\sin \omega x)/\omega$ is a fundamental solution of the operator $Q(d/dx) = (d^2/dx^2) + \omega^2$, ω a real number.

21. Compute $(d/dx)[PV(1/x)]$.

22. Let

$$\mathrm{FP} \int_{-\infty}^{+\infty} \frac{\phi(x)}{x^2}\, dx = \lim_{\varepsilon \to 0} \left[\int_{-\infty}^{-\varepsilon} \frac{\phi(x)}{x^2}\, dx + \int_{\varepsilon}^{+\infty} \frac{\phi(x)}{x^2}\, dx - 2\, \frac{\phi(0)}{\varepsilon} \right]$$

denote the *finite part* of the integral on the left-hand side. Prove that the map

$$\phi \in C_c^\infty(\underline{R}) \to \mathrm{FP} \int_{-\infty}^{+\infty} \frac{\phi(x)}{x^2}\, dx$$

defines a distribution on $\underline{R}$, denoted by $FP(1/x^2)$. Compare it to $(d/dx)[PV(1/x)]$.

23. Prove that the map

$$\phi \in C_c^\infty(\underline{R}) \to FP \int_0^\infty \frac{\phi(x)}{x}\, dx,$$

where

$$FP \int_0^\infty \frac{\phi(x)}{x}\, dx = \lim_{\varepsilon\to 0} \left[\int_\varepsilon^\infty \frac{\phi(x)}{x}\, dx + \phi(0)\, \log \varepsilon \right]$$

defines a distribution on $\underline{R}$, denoted by $FP[Y(x)/x]$. Compute the derivative of $FP[Y(x)/x]$.

24. Prove that the limit

$$\lim_{\varepsilon\to 0} \left[\int_\varepsilon^{+\infty} \frac{\phi(x)}{x}\, dx - \frac{\phi(0)}{\varepsilon} + \phi'(0)\, \log \varepsilon \right]$$

is equal to the *finite part* of the integral $\int_0^\infty [\phi(x)/x^2]\, dx$. Then, define the distribution $FP[Y(x)/x^2]$ and compare it with $(d/dx)\{FP[Y(x)/x]\}$.

Chapter 3

CONVOLUTIONS

1. THE DIRECT PRODUCT OF DISTRIBUTIONS

Let Ω (resp. Ω') be an open subset of $\underline{R}^n$ (resp. $\underline{R}^m$). If $\phi \in C_c^\infty(\Omega)$ and $\psi \in C_c^\infty(\Omega')$ we denote by $\phi \otimes \psi$ the function

$$\phi \otimes \psi(x,y) = \phi(x)\cdot\psi(y).$$

The *algebraic tensor product* $C_c^\infty(\Omega) \otimes C_c^\infty(\Omega')$ is the vector space of all functions $u(x,y)$ that can be represented as finite sums

$$u(x,y) = \sum \phi_j(x)\cdot\psi_j(y)$$

where $\phi_j \in C_c^\infty(\Omega)$ and $\psi_j \in C_c^\infty(\Omega')$. Clearly, $C_c^\infty(\Omega) \otimes C_c^\infty(\Omega')$ is a vector subspace of $C_c^\infty(\Omega \times \Omega')$, the space of all C^∞ functions in the variables $(x,y) = (x_1,\cdots,x_n, y_1,\cdots,y_m)$ with compact support in $\Omega \times \Omega'$. It can be proved that the restrictions to $\Omega \times \Omega'$ of polynomials in (x,y) are dense in $C^\infty(\Omega \times \Omega')$ equipped with its natural topology. (See [17, p. 369].) It then follows that the tensor product $C_c^\infty(\Omega) \otimes C_c^\infty(\Omega')$ is *dense* in $C_c^\infty(\Omega \times \Omega')$. Indeed, if for every $u \in C_c^\infty(\Omega \times \Omega')$ there is a sequence of polynomials $(P_k(x,y))$ converging to u in $C^\infty(\Omega \times \Omega')$ then, taking $\alpha \in C_c^\infty(\Omega)$ and $\beta \in C_c^\infty(\Omega')$ such that $\alpha(x)\beta(y) = 1$ on a neighborhood of the support of u, the sequence $(\alpha(x)\beta(y)P_k(x,y)$ belongs to $C_c^\infty(\Omega) \otimes C_c^\infty(\Omega')$ and converges to u in $C_c^\infty(\Omega \times \Omega')$. As a consequence, it follows that *every distribution* $T \in \mathcal{D}'(\Omega \times \Omega')$ *is well determined by its values on functions* $\phi \otimes \psi$ *where* $\phi \in C_c^\infty(\Omega)$ *and* $\psi \in C_c^\infty(\Omega')$.

Theorem 3.1. *Let* $S \in \mathcal{D}'(\Omega)$ *and* $T \in \mathcal{D}'(\Omega)$. *There is one and only one distribution* $S \otimes T \in \mathcal{D}'(\Omega \times \Omega')$ *defined by*

$$\langle S \otimes T, u(x,y)\rangle = \langle S_x, \langle T_y, u(x,y)\rangle\rangle = \langle T_y, \langle S_x, u(x,y)\rangle\rangle \quad (3.1)$$

for all $u \in C_c^\infty(\Omega \times \Omega')$ *and such that*

$$\langle S \otimes T, \phi \otimes \psi\rangle = \langle S,\phi\rangle \cdot \langle T,\psi\rangle \quad (3.2)$$

for all $\phi \in C_c^\infty(\Omega)$ *and all* $\psi \in C_c^\infty(\Omega')$.

The distribution $S \otimes T$ is said to be the *direct* (or *tensor*) *product* of S and T.

Formula (3.1) can be viewed as an extension of Fubini's theorem. Indeed, when $S = f(x)$ and $T = g(y)$ are two locally integrable functions in Ω and Ω', respectively, we have

$$\langle S_x, \langle T_y, u(x,y)\rangle\rangle = \int_\Omega f(x) \left\{\int_{\Omega'} g(y)u(x,y)\, dy\right\} dx$$

and

$$\langle T_y, \langle S_x, u(x,y)\rangle\rangle = \int_{\Omega'} g(y) \left\{\int_{\Omega} f(x)u(x,y)\, dx\right\} dy$$

which, by Fubini's theorem, are equal to

$$\langle S \otimes T, u\rangle = \int_{\Omega\times\Omega'} f(x)g(y)u(x,y)\, dx\, dy.$$

The proof of Theorem 3.1 uses the following two lemmas which extend to distributions the classical results on continuity and differentiability of integrals with respect to a parameter.

Lemma 3.1. *Let* $\phi(x,\lambda)$ *be a family of functions of* $C_c^\infty(\Omega)$ *depending upon a real or complex parameter* λ. *Suppose that:* (i) *when* $\lambda \in V(\lambda_0)$ *a neighborhood of* λ_0 *the support of* $\phi(x,\lambda)$ *is contained in a fixed compact subset of* Ω; (ii) *for every* $p = (p_1, \cdots, p_n)$ *the derivatives* $(\partial^p \phi/\partial x^p)(x,\lambda)$ *are continuous functions of both variables* (x,λ).

Then

$$\langle T_x, \phi(x, \lambda)\rangle$$

is a continuous function of λ *for every* $T \in \mathcal{D}'(\Omega)$.

Proof. Let

$$\psi_\lambda(x) = \phi(x,\lambda) - \phi(x,\lambda_0).$$

Condition (i) and (ii) imply that $\psi_\lambda \to 0$ in $C_c^\infty(\Omega)$ when $\lambda \to 0$. Hence, for every $T \in \mathcal{D}'(\Omega)$,

$$\langle T_x, \psi_\lambda(x)\rangle \to 0,$$

as $\lambda \to \lambda_0$, which implies the continuity in λ of $\langle T_x, \phi(x,\lambda)\rangle$. Q.E.D

Lemma 3.2. *Let* $\phi(x,\lambda)$ *be a family of functions of* $C_c^\infty(\Omega)$ *depending upon a real or complex parameter. Suppose that:*

(i) *When* $\lambda \in V(\lambda_0)$ *a neighborhood of* λ_0, *the support of* $\phi(x,\lambda)$ *is contained in a fixed compact subset of* Ω.

(ii) *For every* $p = (p_1, \cdots, p_n)$, $(\partial^p \phi/\partial x^p)(x,\lambda)$ *has a derivative (in the classical sense)*

$$\frac{\partial}{\partial\lambda}\frac{\partial^p \phi}{\partial x^p}(x,\lambda)$$

which is continuous in both variables (x,λ).

Then, for every $T \in \mathcal{D}'(\Omega)$,

$$\langle T_x, \phi(x,\lambda)\rangle$$

is differentiable in a neighborhood of λ_0 *and*

$$\frac{\partial}{\partial\lambda}\langle T_x, \phi(x,\lambda)\rangle = \langle T_x, \frac{\partial\phi}{\partial\lambda}(x,\lambda)\rangle.$$

Proof. Let

$$\psi_h(x) = \frac{\phi(x,\lambda+h)-\phi(x,\lambda)}{h} - \frac{\partial\phi}{\partial\lambda}(x,\lambda).$$

It is easy to see that conditions (i) and (ii) imply that $\psi_h \to 0$ in $C_c^\infty(\Omega)$, hence, for every $T \in \mathcal{D}'(\Omega)$,

$$\langle T_x, \frac{\phi(x,\lambda+h) - \phi(x,\lambda)}{h}\rangle \to \langle T_x, \frac{\partial\phi}{\partial\lambda}(x,\lambda)\rangle$$

as $h \to 0$. Q.E.D.

Remarks. 1. Lemmas 3.1 and 3.2 are obviously true in the case of dependence upon multiple parameters $(\lambda_1,\cdots,\lambda_r)$.

2. There are analogies of Lemmas 3.1 and 3.2 in the case where $T \in \mathcal{E}'(\Omega)$ and $\phi(x,\lambda)$ is a family of functions of $C^\infty(\Omega)$.

Proof of Theorem 3.1. 1. Let $u \in C_c^\infty(\Omega \times \Omega')$. By lemma 3.2,

$$\langle T_y, u(x,y)\rangle$$

is a C^∞ function of the variable $x = (x_1,\cdots,x_n)$ and, clearly, it has a compact support in Ω. Thus, we can apply to it the distribution S_x and get the "iterated integral"

$$\langle S_x, \langle T_y, u(x,y)\rangle\rangle. \tag{3.3}$$

In the same way one can see that

$$\langle T_y, \langle S_x, u(x,y)\rangle\rangle \tag{3.4}$$

is well defined.

2. If $u(x,y) = \phi(x)\cdot\psi(y)$ with $\phi \in C_c^\infty(\Omega)$ and $\psi \in C_c^\infty(\Omega')$, it is quite clear that

$$\langle S_x, \langle T_y, u(x,y)\rangle\rangle = \langle T_y, \langle S_x, u(x,y)\rangle\rangle = \langle S,\phi\rangle\cdot\langle T,\psi\rangle,$$

hence the two linear functionals (3.3) and (3.4) coincide on $C_c^\infty(\Omega) \otimes C_c^\infty(\Omega')$.

3. Finally, the linear map

$$u \in C_c^\infty(\Omega \times \Omega) \to \langle T_y, u(x,y)\rangle \in C_c^\infty(\Omega) \tag{3.5}$$

is a continuous one. Indeed, it suffices to show (and this we leave to the reader as an exercise) that it is continuous on $C_c^\infty(\Omega) \otimes C_c^\infty(\Omega')$ equipped with the topology induced by $C_c^\infty(\Omega \times \Omega')$. This result implies that both (3.3) and (3.4) are continuous on $C_c^\infty(\Omega \times \Omega')$. Q.E.D.

Properties of the Direct Product

1. *The support of* $S \otimes T$ *is equal to the cartesian product of the support of* S *by the support of* T.

The proof is easy.

2. *The bilinear map*

$$(S,T) \in \mathcal{D}'(\Omega) \times \mathcal{D}'(\Omega') \to S \otimes T \in \mathcal{D}'(\Omega \times \Omega')$$

is continuous on each variable.

Proof. The continuity with respect to the first variable, say, follows from

$$\langle S \otimes T, u\rangle = \langle S_x, \langle T_y, u(x,y)\rangle\rangle$$

taking into account the continuity of (3.5). Q.E.D.

3. *For every* $p = (p_1,\cdots,p_n)$ *and* $q = (q_1,\cdots,q_n)$ *we have*

$$\frac{\partial^p}{\partial x^p}\frac{\partial^q}{\partial y^q}(S \otimes T) = \frac{\partial^p S}{\partial x^p} \otimes \frac{\partial^q T}{\partial y^q}.$$

The proof is easy.

2. CONVOLUTION OF DISTRIBUTIONS

In Chapter 1, formula (1.1), we have defined the convolution of two functions in a special case, one of the functions being locally integrable and the other being a C^∞ function with compact support. More generally, if f and g are two locally integrable functions in $\underline{R}^n$, one of which has compact support, their convolution $f*g$ is defined by

$$(f*g)(x) = \int_{\underline{R}^n} f(x-y)g(y)\, dy = \int_{\underline{R}^n} f(y)g(x-y)\, dy. \qquad (3.6)$$

In order to extend the definition of convolution to the case of distributions we have to interpret the last formula as a continuous linear functional on $C_c^\infty(\underline{R}^n)$. For every $\phi \in C_c^\infty(\underline{R}^n)$ we have

$$\langle f*g, \phi\rangle = \int_{\underline{R}^n} (f*g)(x)\phi(x)\, dx.$$

Replacing $f*g$ by one of the above integrals and changing variables, we get

$$\langle f*g, \phi\rangle = \int_{\underline{R}^n}\int_{\underline{R}^n} f(x-y)g(y)\phi(x)\, dx\, dy$$

$$= \int_{\underline{R}^n}\int_{\underline{R}^n} f(\xi)g(\eta)\phi(\xi+\eta)\, d\xi\, d\eta$$

$$= \langle f(\xi) \otimes g(\eta), \phi(\xi+\eta)\rangle.$$

Thus, we can redefine the convolution of f and g by the following functional relation:

$$\langle f*g, \phi\rangle = \langle f(\xi) \otimes g(\eta), \phi(\xi+\eta)\rangle, \ \forall\phi \in C_c^\infty(\underline{R}^n).$$

Observe that we still have to give a meaning to the last bracket since $\phi(\xi+\eta)$ as a function of (ξ,η) *does not* have compact support in $R^n \times R^n$. This will be done in a more general situation.

Let $S \in \mathcal{E}'(R^n)$ and $T \in \mathcal{D}'(R^n)$. If $\phi \in C_c^\infty(\underline{R}^n)$ and its support is K, then the support of $\phi(\xi+\eta)$ is contained in the strip

$$\{(\xi,\eta) \in \underline{R}^n \times \underline{R}^n \colon \xi + \eta \in K\}.$$

Since the support of S is a compact set, say L, and the support of $S \otimes T$ is contained in $L \times \underline{R}^n$, by property 1 above it is easy to see that the intersection

$$\operatorname{supp}(S \otimes T) \cap \operatorname{supp} \phi(\xi+\eta)$$

is a compact subset M of $\underline{R}^n \times \underline{R}^n$. Let $\alpha(\xi,\eta)$ be a function of $C_c^\infty(\underline{R}^n \times \underline{R}^n)$ equal to one on M and define

$$\langle S*T, \phi\rangle = \langle S_\xi \otimes T_\eta, \phi(\xi+\eta)\rangle = \langle S_\xi \otimes T_\eta, \alpha(\xi,\eta)\phi(\xi+\eta)\rangle.$$

Observe that the last bracket is now well defined because $\alpha(\xi,\eta)\phi(\xi+\eta)$ has compact support. Moreover, its value is independent of the choice of α because α is taken equal to one on M. This discussion justifies the following definition.

Definition 3.1. *Let* $S,T \in \mathcal{D}'(\underline{R}^n)$ *and suppose that at least one has compact support. The convolution* $S*T$ *is a distribution in* $\underline{R}^n$ *defined by*

$$\langle S*T, \phi\rangle = \langle S_\xi \otimes T_\eta, \phi(\xi+\eta)\rangle, \ \forall\phi \in C_c^\infty(\underline{R}^n). \tag{3.7}$$

It is quite clear that this new definition coincides with the classical one (3.6) when $S = f$ and $T = g$ are locally integrable functions and at least one has compact support.

As we shall see later, the condition that at least one of the distributions must have compact support can be lifted in many cases.

Roughly speaking, when the growth at infinity of one of the distributions is compensated by the decay of the other, the convolution product is well defined.

We shall mention for future reference the following important results on the convolution of functions:

(I) *Let* p, q, *and* r *be real numbers such that* $1 \leq p,q,r \leq +\infty$ *and* $r^{-1} = p^{-1} + q^{-1} - 1$. *If* $f \in L^p(\underline{R}^n)$ *and* $g \in L^q(\underline{R}^n)$, *then* $f*g \in L^r(\underline{R}^n)$ *and*

$$\|f*g\|_r \leq \|f\|_p \cdot \|g\|_q .$$

For its proof, the reader should consult Treves [32, p. 278] or Zygmund [33]. As a consequence, we have the following:

(II) *Let* p = 1. *For every* $f \in L^1$, *the map*

$$g \in L^q \to f*g \in L^p$$

is a continuous linear one whose norm is $\leq \|f\|$.

(III) *The bilinear map*

$$L^1 \times L^1 \ni (f,g) \to f*g \in L^1$$

is continuous and we have $\|f*g\|_1 \leq \|f\|_1 \cdot \|g\|_1$.

Properties of the Convolution

1. *Let* $S,T \in \mathcal{D}'(\underline{R}^n)$ *and suppose that at least one has compact support. We have*

$$\text{supp}(S*T) \subset \text{supp } S + \text{supp } T.$$

Proof. Let A = supp S and B = supp T. Since A and B are closed and at least one is compact, the set

$$A + B = \{x + y: x \in A, y \in B\}$$

is *closed*. Let us show that $S*T$ is equal to zero on the open set $\Omega = (A + B)^c$. Indeed, if $\phi \in C_c^\infty(\Omega)$, the support of $\phi(\xi + \eta)$ is contained in the open set

$$\{(\xi,\eta) \in \underline{R}^n \times \underline{R}^n: \xi + \eta \in \Omega\}.$$

On the other hand, the support of $S \otimes T$ is, as we have seen, $A \times B$. Since $(\xi,\eta) \in A \times B$ implies $\xi + \eta \in A + B$, the support of $S \otimes T$ does not intersect the support of $\phi(\xi + \eta)$; consequently

$$<S*T, \phi> = <S_\xi \otimes T_\eta, \phi(\xi + \eta)>,$$

for all $\phi \in C_c^\infty(\Omega)$. Q.E.D.

2. *The bilinear map*

$$(S,T) \in \mathcal{E}'(\underline{R}^n) \times \mathcal{D}'(\underline{R}^n) \to S*T \in \mathcal{D}'(\underline{R}^n)$$

is continuous on each variable.

The proof, a consequence of property 2 of the direct product is left to the reader.

3. $\mathcal{E}'(\underline{R}^n)$ *is a commutative and associative algebra with unit element with respect to the convolution product.*

Proof. The convolution product is obviously a *commutative* one.

The Dirac measure δ is the *unit element* of this product. In fact,

$$<\delta*T, \phi> = <\delta_\xi \otimes T_\eta, \phi(\xi + \eta)$$

$$= <T_\eta, <\delta_\xi, \phi(\xi + \eta)>> = <T ,\phi>$$

for all $\phi \in C_c^\infty(\underline{R}^n)$.

Let R, S, T $\in \mathcal{D}'(\underline{R}^n)$ and suppose that *at least two of them* have compact support. Then, the convolution product is associative and we have

$$R*S*T = R*(S*T) = (R*S)*T. \tag{3.9}$$

Indeed, from our assumption about the supports of R, S, T, it follows that both sides of (3.9) are well defined. Also, if ϕ is an arbitrary element of $C_c^\infty(\underline{R}^n)$ it is easy to check by applying both sides of (3.9) to ϕ that we get the same value

$$\langle R_\xi \otimes S_\eta \otimes T_\zeta, \phi(\xi + \eta + \zeta)\rangle.$$

4. *Convolutions and translations.* Let h be an element of $\underline{R}^n$ and let ϕ be a function defined in $\underline{R}^n$. The *translation* of ϕ by h is the function defined by

$$(\tau_h\phi)(x) = \phi(x - h).$$

It is obvious that if $\phi \in C^\infty$ (resp. C_c) then $\tau_h\phi \in C^\infty$ (resp. C_c^∞). We define the translation of a distribution by duality:

$$\langle\tau_h T,\phi\rangle = \langle T,\tau_{-h}\phi\rangle, \ \forall\phi \in C_c^\infty(\underline{R}^n).$$

As an exercise, the reader can show that for every h $\in \underline{R}^n$, $\tau_h T$ is a distribution and

$$\tau_h: \mathcal{D}'(\underline{R}^n) \to \mathcal{D}'(\underline{R}^n)$$

is a continuous linear map in the strong topologies.

Let us denote by $\delta_{(h)}$ *the Dirac measure at the point* h $\in \underline{R}^n$, i.e.,

$$<\delta_{(h)}, \phi> = \phi(h), \forall\phi \in C_c^\infty(\underline{R}^n).$$

(We set $\delta_{(0)} = \delta$.) Then, the following property holds true:

$$\tau_h T = \delta_{(h)}*T, \forall T \in \mathcal{D}'(\underline{R}^n). \tag{3.10}$$

Indeed, we have

$$<\tau_h T, \phi> = <T, \tau_{-h}\phi> = <T_\xi, \phi(\xi + h)>.$$

Replacing $\phi(\xi + h)$ by

$$\phi(\xi + h) = <(\delta_{(h)})_\eta, \phi(\xi + \eta)>$$

we get

$$<\tau_h T, \phi> = <T_\xi, <(\delta_{(h)})_\eta, \phi(\xi + \eta)>>$$

$$= <T_\xi \otimes (\delta_{(h)})_\eta, \phi(\xi + \eta)> = <\delta_{(h)}*T, \phi> . \quad \text{Q.E.D.}$$

By using the commutative and associative properties of the convolution product we can derive, as a consequence of (3.10), the following very useful formula

$$\tau_h(S*T) = (\tau_h S)*T = S*(\tau_h T). \tag{3.11}$$

5. *The derivative as a convolution product.* The following relation holds:

$$\partial_k T = \partial_k\delta*T. \tag{3.12}$$

Indeed, we have

$$<\partial_k T, \phi> = -<T, \partial_k\phi>.$$

By writing

$$\partial_k\phi(x) = \langle\delta_y, \partial_k\phi(x + y)\rangle = -\langle(\partial_k\delta)_y, \phi(x + y)\rangle$$

and making the appropriate replacement, we get

$$\langle\partial_k T, \phi\rangle = -\langle T_x, \partial_k\phi(x)\rangle = \langle T_x, \langle(\partial_k\delta)_y, \phi(x + y)\rangle\rangle$$

$$= \langle T_x \otimes (\partial_k\delta)_y, \phi(x + y)\rangle = \langle(\partial_k\delta)*T,\phi\rangle ,$$

for all $\phi \in C_c^\infty(\underline{R}^n)$. Q.E.D.

Formula (3.12) implies (using the associativity of the convolution) that *in order to differentiate a convolution product it suffices to differentiate one of the factors*, i.e.,

$$\partial_k(S*T) = \partial_k S*T = S*\partial_k T. \tag{3.13}$$

3. CONVOLUTION OF FUNCTIONS AND DISTRIBUTIONS; REGULARIZATION

Let $\phi \in C_c^\infty(\underline{R}^n)$ [resp. $\phi \in C^\infty(\underline{R}^n)$] and let $T \in \mathcal{D}'(\underline{R}^n)$ [resp. $\phi \in \mathcal{E}'(\underline{R}^n)$]. Then the convolution of T and ϕ is given by the formula

$$(T*\phi)(x) = \langle T_\xi, \phi(x - \xi)\rangle. \tag{3.14}$$

In fact, for every $\psi \in C_c^\infty(\underline{R}^n)$ we have, by (3.7),

$$\langle(T*\phi)(x), \psi(x)\rangle = \langle T_\xi \otimes \phi(\eta), \psi(\xi + \eta)\rangle = \langle T_\xi, \langle\phi(\eta), \psi(\xi + \eta)\rangle\rangle.$$

But

$$\langle\phi(\eta), \psi(\xi + \eta)\rangle = \int_{\underline{R}^n} \phi(\eta)\psi(\xi + \eta)\, d\eta$$

$$= \int_{\underline{R}^n} \phi(x - \xi)\psi(x)\, dx = \langle\phi(x - \xi), \psi(x)\rangle.$$

Making the appropriate replacement, we get

$$<(T*\phi)(x), \psi(x)> = <T_\xi, <\phi(x - \xi), \psi(x)>>$$

$$= <<T_\xi, \phi(x - \xi)>, \psi(x)>. \quad \text{Q.E.D.}$$

Let us set

$$\check{\phi}(x) = \phi(-x).$$

Using this notation, (3.14) can be written as

$$(T*\phi)(x) = <T, \tau_x\check{\phi}> \tag{3.15}$$

We also have

$$<T, \phi> = (T*\check{\phi})(0). \tag{3.16}$$

Theorem 3.2. Let $\phi \in C_c^\infty(\underline{R}^n)$ *[resp.* $\phi \in C^\infty(\underline{R}^n)$*] and let* $T \in \mathcal{D}'(\underline{R}^n)$ *[resp.* $T \in \mathcal{E}'(\underline{R}^n)$*]. Then:*

1. $(T*\phi)(x) \in C^\infty(\underline{R}^n)$.
2. *The bilinear map*

$$(\phi, T) \to T*\phi$$

from $C_c^\infty(\underline{R}^n) \times \mathcal{D}'(\underline{R}^n)$ *[resp.* $C^\infty(\underline{R}^n) \times \mathcal{E}'(\underline{R}^n)$*] into* $C^\infty(\underline{R}^n)$ *is separately continuous.*

Proof. 1. Since

$$(T*\phi)(x) = <T_y, \phi(x - y)>$$

then, by Lemma 3.1, it follows that $(T*\phi)(x)$ is a continuous function of x. Taking derivatives and using (3.13) we get that, for all $p = (p_1,\cdots,p_n)$,

$$\partial^p(T*\phi) = T*\partial^p\phi$$

is a continuous function; therefore, $T*\phi \in C^\infty$.

2. To show that the map

$$\phi \in C_c^\infty \to T*\phi \in C^\infty$$

is a continuous one, it suffices to show by Proposition 1.1 that for every compact subset $K \subset \underline{R}^n$, the map

$$\phi \in C_c^\infty(\underline{R}^n;K) \to T*\phi \in C^\infty(\underline{R}^n)$$

is a continuous one. In other words, for all $p \in \underline{N}^n$, for all L a compact subset of $\underline{R}^n$, there is a constant $C > 0$ and an integer $m > 0$ such that

$$\sup_{x\in L}|\partial^p(T*\phi)(x)| \le C \sup_{\substack{|q|\le m\\ z\in K}} |\partial^q\phi(z)|.$$

But when $x \in L$ and $y \in K$, it is clear that $(\tau_y\phi)(x) = \phi(x - y)$ belongs to $C_c^\infty(\underline{R}^n;L - K)$. Since $T \in \mathcal{D}'$, by Theorem 2.12, there is a constant $C > 0$ and an integer $\ell \ge 0$ such that

$$|\langle T, \psi\rangle| \le C \sup_{\substack{|r|\le\ell\\ y\in\underline{R}^n}} |\partial^r\psi(y)|, \forall\ \psi \in C_c^\infty(\underline{R}^n;\ L - K).$$

Since $\partial^p(T*\phi)(x) = (T*\partial^p\phi)(x) = \langle T_y, \partial^p\phi(x - y)\rangle$, we get

$$\sup_{x\in L} |\partial^p(T*\phi)(x)| = \sup_{x\in L}|\langle T_y, \partial^p\phi(x - y)\rangle|$$

$$\le C \sup_{x\in L} \sup_{\substack{|r|\le\ell\\ y\in\underline{R}^n}} |\partial_y^r\partial_x^p\phi(x - y)| \le C \sup_{\substack{z\in\underline{R}^n\\ |q|\le\ell+|p}} |\partial^q\phi(z)|,$$

as we wanted to prove.

3. Let ϕ be a fixed element of $C_c^\infty(\underline{R}^n)$. It is easy to see that if x belongs to a compact subset K of $\underline{R}^n$ the set of functions $\{\tau_x\check{\phi}: x \in K\}$ is bounded in $C_c^\infty(\underline{R}^n)$. Hence, if the distributions T_j converge strongly to zero, then

$$(T_j*\phi)(x) = \langle T_j, \tau_x\check{\phi}\rangle$$

converge to zero uniformly on K. By taking derivatives and using (3.13) we also get that for all $p \in \underline{N}^n$ the functions

$$\partial^p(T_j*\phi)(x) = (T_j*\partial^p\phi)(x)$$

converge to zero uniformly on K an arbitrary compact subset of $\underline{R}^n$. Therefore $T_j*\phi \to 0$ in $C^\infty(\underline{R}^n)$. Q.E.D.

Remark. With a proof similar to that of Theorem 3.2, one can show that if $T \in \mathcal{E}'(\underline{R}^n)$ and $\phi \in C_c(\underline{R}^n)$ then $T*\phi \in C_c^\infty(\underline{R}^n)$ and the bilinear map

$$(T,\phi) \in \mathcal{E}'(\underline{R}^n) \times C_c^\infty(\underline{R}^n) \to T*\phi \in C_c^\infty(\underline{R}^n)$$

is continuous on each variable.

Regularization of Distribution

Let

$$\alpha_\varepsilon(x) = \varepsilon^{-n}\,\alpha\left(-\frac{x}{\varepsilon}\right)$$

be the test function considered in Chapter 1, Section 2. We have the following theorem.

Theorem 3.3. *If* $T \in \mathcal{D}'(\underline{R}^n)$ [*resp.* $T \in \mathcal{E}'(\underline{R}^n)$], *the functions* $T*\alpha_\varepsilon$ *of* $C^\infty(\underline{R}^n)$ [*resp.* $C_c^\infty(\underline{R}^n)$] *converge strongly to* T *in* $\mathcal{D}'(\underline{R}^n)$. [*resp.* $\mathcal{E}'(\underline{R}^n)$] *when* $\varepsilon \to 0$.

The proof is based upon the following lemma.

Lemma 3.4. *If* $\psi \in C_c^\infty(\underline{R}^n)$, *the functions* $\alpha_\varepsilon * \psi$ *of* $C_c^\infty(\underline{R}^n)$ *converge to* ψ *in* $C_c^\infty(\underline{R}^n)$ *when* $\varepsilon \to 0$. *Moreover, the convergence is uniform on bounded sets of* $C_c^\infty(\underline{R}^n)$.

Proof. Let B be a bounded set in $C_c^\infty(\underline{R}^n)$. As we know (Theorem 1.3), there is a compact subset L of $\underline{R}^n$ such that $\operatorname{supp} \psi \subset L$, $\forall \psi \in B$, and such that B is a bounded subset of $C_c^\infty(\underline{R}^n;L)$. On the other hand, by Theorem 1.1, property (2), there is a compact subset K of $\underline{R}^n$ such that $L \subset K$ and such that

$$\operatorname{supp}(\alpha_\varepsilon * \psi) \subset K$$

for all $\psi \in B$ and $0 < \varepsilon < 1$.

Since, by assumption, B is bounded in $C_c^\infty(\underline{R}^n)$, we can prove as in Lemma 2.1 that for every $p = (p_1, \cdots, p_n)$ the set

$$\partial^p B = \{\partial^p \psi : \psi \in B\}$$

is uniformly equicontinuous on K. Hence, given m a nonnegative integer and $\sigma > 0$ a real number, there is $\delta > 0$ such that

$$|\partial^p \psi(x - y) - \partial^p \psi(x)| \leq \sigma$$

$\forall |y| \leq \delta, \forall \psi \in B, \forall\, x \in K$, and $\forall |p| \leq m$.

Consider now the neighborhood of zero

$$W(m,\sigma) = \{\phi \in C_c^\infty(\underline{R}^n;K): \ |\partial^p \phi(x)| \leq \sigma, \forall\, x \in K,\ \forall |p| \leq m\}$$

in $C_c^\infty(\underline{R}^n;K)$. We can write

$$(\alpha_\varepsilon * \partial^p \psi - \partial^p \psi)(x) = \int_{\underline{R}^n} \alpha_\varepsilon(y)(\partial^p \psi(x - y) - \partial^p \psi(x))\, dy.$$

Let $\varepsilon_0 > 0$ be small enough so that for all $\varepsilon < \varepsilon_0$ the function α_ε has its support contained in the ball of center at the origin and of radius δ. It follows that

$$|(\alpha_\varepsilon * \partial^p\psi - \partial^p\psi)(x)| \le \int_{\underline{R}^n} \alpha_\varepsilon(y)|\partial^p\psi(x - y) - \partial^p\psi(x)|\, dy \le \sigma$$

for all $\psi \in B$, $|p| \le m$, and $\varepsilon \le \varepsilon_0$. In other words,

$$\alpha_\varepsilon * \psi - \psi \in W(m,\sigma),\ \forall\varepsilon \le \varepsilon_0,\ \forall\ \psi \in B$$

which proves that $\alpha_\varepsilon * \psi \to \psi$, when $\varepsilon \to 0$, uniformly on bounded sets of $C_c^\infty(\underline{R}^n)$. Q.E.D.

Remark. Lemma 3.4 holds true if we replace $C_c^\infty(\underline{R}^n)$ by $C^\infty(\underline{R}^n)$.

Proof of Theorem 3.3. Let T be a fixed element of $\mathcal{D}'(\underline{R}^n)$. By (3.16) we have

$$\begin{aligned} \langle T*\alpha_\varepsilon - T, \psi\rangle &= [(T*\alpha_\varepsilon)*\check{\psi} - T*\check{\psi}](0) \\ &= [T*(\alpha_\varepsilon*\check{\psi}) - T*\check{\psi}](0) = [T*(\alpha_\varepsilon*\check{\psi} - \check{\psi})](0) \\ &= \langle T, (\check{\alpha}_\varepsilon*\psi - \psi)^{\vee}\rangle. \end{aligned}$$

By Lemma 3.4, $\check{\alpha}_\varepsilon*\psi - \psi \to 0$ uniformly when ψ belongs to a bounded subset of $C_c^\infty(\underline{R}^n)$, as $\varepsilon \to 0$. This implies that

$$\langle T, (\check{\alpha}_\varepsilon*\psi - \psi)^{\vee}\rangle \to 0$$

and therefore that $T*\alpha_\varepsilon$ converges strongly to T in $\mathcal{D}'(\underline{R}^n)$ when $\varepsilon \to 0$. The same proof applies to the case when $T \in \mathcal{E}'(\underline{R}^n)$. Q.E.D.

Corollary. The family of functions α_ε converges to the Dirac measure δ in the strong topology of $\mathcal{D}'(\underline{R}^n)$.

Proof. It suffices to apply Theorem 3.3 to $T = \delta$, recalling that δ is the unit element with respect to the convolution product. Q.E.D.

Remark. Since $\delta \in \mathcal{E}'(\underline{R}^n)$ we also have $\alpha_\varepsilon \to \delta$ in the strong topology of $\mathcal{E}'(\underline{R}^n)$ when $\varepsilon \to 0$.

According to Definition 1.5 of Chapter 1, we called (α_ε) a regularizing family of functions and

$$\alpha_j(x) = j^n \alpha(jx), \ j = 1,2,\cdots,$$

a regularizing sequence of functions. In Theorem 1.1, we proved that integrable functions can be approximated in suitable topologies by smooth functions. Theorem 3.3 extends this result to distributions; it allows us to replace a distribution (which in general has singularities) by a converging family (or sequence) of functions.

Convolution maps

Let $T \in \mathcal{D}'(\underline{R}^n)$. The continuous linear map

$$L_T: \phi \in C_c^\infty(\underline{R}^n) \to T*\phi \in C^\infty(\underline{R}^n)$$

is said to be a *convolution map*. To every $T \in \mathcal{D}'(\underline{R}^n)$ we can then associate a convolution map. The following theorem characterizes such maps.

Theorem 3.4. *Let* $T \in \mathcal{D}'(\underline{R}^n)$. *The convolution map* L_T *is a continuous linear map from* $C_c^\infty(\underline{R}^n)$ *into* $C^\infty(\underline{R}^n)$ *which commutes with translations.*

Conversely, if $L: C_c^\infty(\underline{R}^n) \to C^\infty(\underline{R}^n)$ *is a continuous linear map such that*

$$L \circ \tau_h = \tau_h \circ L,$$

for every $h \in \underline{R}^n$, *there is a unique* $T \in \mathcal{D}'(\underline{R}^n)$ *such that*

$$L(\phi) = T*\phi, \forall\phi \in C_c^\infty(\underline{R}^n).$$

Proof. Let $T \in \mathcal{D}'(\underline{R}^n)$. By (3.14) we have

$$[T*(\tau_h\phi)](x) = \langle T_y, (\tau_h\phi)(x - y)\rangle = \langle T_y, \phi(x - y - h)\rangle$$

$$= (T*\phi)(x - h) = [\tau_h(T*\phi)](x)$$

which proves that convolution maps commute with translations.

Conversely, suppose that $L: C_c(\underline{R}^n) \to C^\infty(\underline{R}^n)$ is a continuous linear map commuting with translations. It is easy to see that the map

$$\phi \in C_c^\infty(\underline{R}^n) \to (L\phi)(0) \in \underline{C}$$

defines a continuous linear functional on $C_c^\infty(\underline{R}^n)$; hence, there is a unique $T \in \mathcal{D}'(\underline{R}^n)$ such that

$$(L\phi)(0) = \langle T, \check{\phi}\rangle.$$

By (3.16) we get

$$(L\phi)(0) = (T*\phi)(0).$$

Since L commutes with translations we have for every $x \in \underline{R}^n$

$$(L\phi)(x) = [\tau_{-x}(L\phi)](0) = [L(\tau_{-x}\phi)](0) = (T*(\tau_{-x}\phi)(0)$$

$$=\langle T_y, (\tau_{-x}\phi)(-y)\rangle = \langle T_y, \phi(x - y)\rangle = (T*\phi)(x)$$

which shows that L is a convolution map. Q.E.D.

Remark. If $T \in \mathcal{E}'(\underline{R}^n)$, it defines a convolution map from $C^\infty(\underline{R}^n)$ into $C^\infty(\underline{R}^n)$. In this case, we also have a characterization analogous to that of Theorem 3.4.

PROBLEMS

1. Prove that the family of functions (ψ_λ) introduced in the proof of Lemma 3.1 converges to zero in $C_c^\infty(\underline{R}^n)$ as $\lambda \to \lambda_0$.

2. Answer Problem 1 for the family introduced in the proof of Lemma 3.2.

3. Prove that if $T \in \mathcal{D}'(\Omega \times \Omega')$ [resp. $\mathcal{E}'(\Omega \times \Omega')$], the linear map

$$u \in C_c^\infty(\Omega \times \Omega') \text{ [resp. } C^\infty(\Omega \times \Omega')] \to \langle T_y, u(x,y)\rangle \in C_c^\infty(\Omega) \text{ [resp. } C^\infty(\Omega)]$$

is a continuous one.

4. Prove properties 1 and 3 of the direct product of distributions.

5. Prove the continuity on each variable of the bilinear map

$$(S,T) \in \mathcal{E}'(\underline{R}^n) \times \mathcal{D}'(\underline{R}^n) \to S*T \in \mathcal{D}'(\underline{R}^n).$$

6. Prove that τ_h is an isomorphism from $C_c^\infty(\underline{R}^n)$ [resp. $C^\infty(\underline{R}^n)$] onto $C_c^\infty(\underline{R}n)$ [resp. $C^\infty(\underline{R}^n)$].

7. Prove formula (3.13).

8. Prove Theorem 3.2 in the case $\phi \in C^\infty$ and $T \in \mathcal{E}'$.

9. If B is a bounded set in $C^1(\Omega)$, show that B is uniformly equicontinuous on every compact subset of Ω.

10. State and prove the corresponding version of Lemma 3.1 for Radon measures.

11. Let $\mu \in M(\underline{R}^n)$ be a Radon measure on $\underline{R}^n$ and let $\phi \in C_c(\underline{R}^n)$. Prove that

$$(\mu*\phi)(x) = \int_{\underline{R}^n} \phi(x - y)\, d\mu\,(y)$$

is a continuous function in $\underline{R}^n$.

12. If P(x) is a polynomial of degree $\leq m$ and T is a distribution with compact support, prove that P*T is also a polynomial of degree $\leq m$.

13. Consider the following distributions on the real line: $R = 1$, $S = \delta'$, and $T = Y(x)$, the Heaviside function. Prove that in this case formula (3.9) does not hold.

14. Let $\chi = \chi(x)$ be the characteristic function of the interval $(-1,1)$. Compute the convolution $\chi^{(*n)} = \chi*\chi*\cdots*\chi$ (n times) and find its support.

15. Let f and g be two locally integrable functions on $\underline{R}$ vanishing on $\underline{R}_- = \{x \in \underline{R}: x \leq 0\}$. Prove that: (i) the convolution $f*g$ is well defined by

$$(f*g)(x) = \int_0^\infty f(x - y)g(y)\, dy;$$

(ii) $f*g$ vanishes on $\underline{R}_-$; (iii) $f*g$ is a locally integrable function on $\underline{R}$.

Chapter 4

TEMPERED DISTRIBUTIONS AND THEIR FOURIER TRANSFORMS

1. THE SPACE OF INFINITELY DIFFERENTIABLE FUNCTIONS RAPIDLY DECREASING AT INFINITY

Definition 4.1. *We say that a function* ϕ *belonging to* $C^\infty(\underline{R}^n)$ *is rapidly decreasing at infinity if*

$$\lim_{|x|\to+\infty} |x^\alpha \partial^p \phi(x)| = 0 \qquad (4.1)$$

for all $\alpha \in \underline{N}^n$ *and all* $p \in \underline{N}^n$.

The set of all C^∞ *functions rapidly decreasing at infinity is a vector space over* $\underline{C}$ *denoted by* $S(\underline{R}^n)$.

It is easy to check that the condition (4.1) is equivalent to each one of the following conditions:

for all $\alpha, p \in \underline{N}^n$, $x^\alpha \partial^p \phi(x)$ *is uniformly bounded on* $\underline{R}^n$. (4.2)

for all integer $k \geq 0$ *and for all* $p \in \underline{N}^n$, $(1+r^2)^{k/2}\partial^p\phi(x)$, *where* $r = |x|$, *is uniformly bounded on* $\underline{R}^n$,

Examples. 1. The space $C_c^\infty(\underline{R}^n)$ is a vector subspace of $S(\underline{R}^n)$.

2. The function $\exp(-|x|^2/2)$ belongs to $S(\underline{R}^n)$.

3. Let $\alpha \in C_c^\infty(\underline{R}^n)$ be such that $0 \leq \alpha \leq 1$, $\operatorname{supp} \alpha \subset B_1$, and $\alpha(0) = 1$. Let (x_j) be a sequence of elements of $\underline{R}^n$ such that $|x_j| + 2 \leq |x_{j+1}|$. Define

$$\gamma(x) = \sum_{j=1}^{\infty} \frac{\alpha(x-x_j)}{(1+|x_j|^2)^j}.$$

The sum is well defined since the supports of the functions $\alpha(x - x_j)$ are disjoint. If $k \in \underline{N}$ and $p \in \underline{N}^n$, we have

$$(1+|x|^2)^k \partial^p \gamma(x) = \frac{(1+|x|^2)^k \quad \partial^p \alpha(x-x_j)}{(1+|x_j|^2)^k \; (1+|x_j|^2)^{j-k}},$$

whenever $|x_j| - 1 \le |x| \le |x_j| + 1$. On the other hand, $(1+|x|^2)/(1+|x_j|^2) \le C$, with a suitable constant, and

$$\sup_{x \in \underline{R}^n} |\partial^p \alpha(x-x_j)| = \sup_{x \in \underline{R}^n} |\partial^p \alpha(x)|.$$

It then follows that $\gamma \in S(\underline{R}^n)$.

Proposition 4.1. *Let* P(x) *be a polynomial with constant coefficients and* $Q(\partial)$ *a partial differential operator with constant coefficients. The following are equivalent conditions:* (1) $\phi \in S$ (2) $\forall P(x)$ and $\forall Q(\partial)$, $P(x)Q(\partial)\phi \in S$; (3) $\forall P(x)$ and $Q(\partial)$, $\forall Q(\partial)$ $(P(x)\phi(x)) \in S$.

Proof. Since $P(x)Q(\partial)\phi$ can be written as a linear combination of terms of the form (4.2), then Condition 1 clearly implies Condition 2. On the other hand, it is obvious that Condition 2 implies Condition 1.

By Leibniz's formula (1.1), it follows that $Q(\partial)(P(x)\phi(x))$ is a linear combination of terms of the form (4.2), hence Condition 1 implies Condition 3. Finally, we leave to the reader the proof that Condition 3 implies Condition 1. Q.E.D.

2. TEMPERED DISTRIBUTIONS

Define on $S(\underline{R}^n)$ the seminorms

$$\gamma_{\alpha,p}(\phi) = \sup_{x \in \underline{R}^n} |x^\alpha \partial^p \phi(x)|, \; \alpha, \; p \in \underline{N}^n.$$

The countable family of seminorms $(\gamma_{\alpha,p})$ defines a Hausdorff locally convex topology on $S(\underline{R}^n)$ which can be proved to be a metrizable and complete topology. Thus $S(\underline{R}^n)$ is a Frechet space. In view of the equivalence between (4.2) and (4.3), the topology of $S(\underline{R}^n)$ can also be defined by the sequence of seminorms

$$\gamma_{k,m}(\phi) = \sup_{|p|\leq m} \sup_{x\in\underline{R}^n} |(1+r^2)^{k/2}\partial^p\phi(x)|, \quad k,m \in \underline{N}.$$

S is a *Montel space*. Indeed, if B is a bounded set of S, is a bounded set in C^∞, because the imbedding $S \to C^\infty$ is continuous (Theorem 4.2). Since C^∞ is a Montel space, B is then a relatively compact set in C^∞. In order to prove that B is relatively compact in S, it suffices to show that if (ϕ_j), a sequence of elements of B, converges to ϕ in C^∞, then $\phi \in S$ and $\phi_j \to \phi$ in S. Since B is bounded in S, then, for all $k \in \underline{N}$ and for all $p \in \underline{N}^n$, there is a constant $C_{k,p}$ such that

$$\sup_{x\in\underline{R}^n} |(1+r^2)^{k/2}\partial^p f(x)| \leq C_{k,p}, \quad f \in B.$$

But this inequality implies that, given $\varepsilon > 0$, there is a constant $M > 0$ such that

$$|(1+r^2)^{k/2}\partial^p f(x)| \leq \varepsilon, \quad r = |x| > M, \quad f \in B.$$

Since $\phi_j \to \phi$ in C^∞, the last inequality implies that

$$|(1+r^2)^{k/2}\partial^p\phi(x)| \leq \varepsilon, \quad r > M,$$

hence $\phi \in S$. On the other hand, since $\phi_j \to \phi$ in C^∞, then $(\partial^p\phi_j)$ converges uniformly to $\partial^p\phi$ on the compact set $\{x \in \underline{R}^n\colon |x| \leq M\}$. But this implies that, given $\varepsilon > 0$, we can find an integer j_0 such that

$$(1+r^2)^{k/2} | \partial^p \phi_j(x) - \partial^p \phi(x)| \leq \varepsilon,$$

for all $|x| \leq M$ and for all $j \geq j_0$. Our last three inequalities imply that

$$\sup_{x \in \underline{R}^n} (1+r^2)^{k/2} |\partial^p \phi_j(x) - \partial^p \phi(x)| \leq \varepsilon$$

for all $j \geq j_0$; therefore, $\phi_j \to \phi$ in S. Q.E.D.

As a consequence, it follows that S is a *reflexive space*.

Theorem 4.1. *A sequence* (ϕ_j) *converges to zero in* S *if and only if one of the following equivalent conditions holds true*:

1. *For all* $\alpha, p \in \underline{N}^n$

$$x^\alpha \partial^p \phi_j(x) \to 0$$

uniformly on $\underline{R}^n$ *as* $j \to +\infty$.

2. *For all polynomials* $P(x)$ *with constant coefficients and for all partial differential operators with constant coefficients* $Q(\partial)$,

$$P(x) Q(\partial) \phi_j \to 0$$

uniformly on $\underline{R}^n$ *as* $j \to +\infty$.

3. *For all* $P(x)$ *and all* $Q(\partial)$ *as before*,

$$Q(\partial)(P(x)\phi_j(x)) \to 0$$

uniformly on $\underline{R}^n$ *as* $j \to +\infty$.

The proof is an easy consequence of the definition of the seminorms $\gamma_{\alpha,p}$ and Proposition 4.1.

Theorem 4.2. *We have the following inclusions*

$$C_c(\underline{R}^n) \subset (\underline{R}^n) \subset C^\infty(\underline{R}^n).$$

with continuous imbeddings. Moreover, $C_c(\underline{R}^n)$ *is a dense subspace of* $S(\underline{R}^n)$ *and* $S(\underline{R}^n)$ *a dense subspace of* $C^\infty(\underline{R}^n)$.

Proof. We have seen (Theorem 2.16) that $C_c^\infty(\underline{R}^n)$ is dense in $C^\infty(\underline{R}^n)$. This implies that $S(\underline{R}^n)$ is dense in $C^\infty(\underline{R}^n)$.

Let $\beta_j \in C_c^\infty(\underline{R}^n)$ be such that $\beta_j = 1$ on the closed ball with center at the origin and radius j. If $\phi \in S(\underline{R}^n)$, the sequence $(\beta_j\phi)$ of functions of $C_c^\infty(\underline{R}^n)$ converges to ϕ *in* $S(\underline{R}^n)$.

In order to prove that the identity map from $C_c^\infty(\underline{R}^n)$ into $S(\underline{R}^n)$ is continuous it suffices to show (Proposition 1.1) that for every compact subset K of $\underline{R}^n$ the identity map from $C_c^\infty(\underline{R}^n; K)$ into $S(\underline{R}^n)$ is a continuous one. Since they are both metrizable spaces, it suffices to show that every sequence (ϕ_j) converging to zero in $C_c^\infty(\underline{R}^n; K)$ converges to zero in $S(\underline{R}^n)$. If $(\phi_j) \to 0$ in $C_c^\infty(\underline{R}^n; K)$ as $j \to +\infty$, then for all $p \in \underline{N}^n$

$$\partial^p\phi_j(x) \to 0 \text{ uniformly on } K \text{ as } j \to +\infty.$$

Since the support of every ϕ_j is contained in K, this implies that, $\forall\alpha$, $p \in \underline{N}^n$, the sequence

$$(x^\alpha\ \partial^p\phi_j)$$

converges to zero uniformly on $\underline{R}^n$ as $j \to +\infty$.

Finally, suppose that (ϕ_j) is a sequence converging to zero in $S(\underline{R}^n)$. This implies that for every $p \in \underline{N}^n$ the sequence $(\partial^p\phi_j)$ converges to zero uniformly on $\underline{R}^n$. In particular, it converges to zero uniformly on every compact subset of $\underline{R}^n$. Q.E.D.

On the other hand, it is easy to see that $S(\underline{R}^n) \subset \mathcal{D}'(\underline{R}^n)$ with continuous imbedding. By Definition 2.9, $S(\underline{R}^n)$ is then a normal space of distributions, hence its dual $S'(\underline{R}^n)$ is a subspace of $\mathcal{D}'(\underline{R}^n)$.

Definition 4.2. *The elements of* $S'(\underline{R}^n)$ *are said to be tempered distributions.*

Examples. 1. As an immediate consequence of Theorem 4.2 we get the following continuous imbeddings $E'(\underline{R}^n) \to S'(\underline{R}^n) \to \mathcal{D}'(\underline{R}^n)$ in the sense of the strong topologies. The first imbedding shows that every distribution with compact support is a tempered one and that the space of all distributions with compact support is a subspace of $S'(\underline{R}^n)$.

2. Every function $f \in L^p(\underline{R}^n)$, $1 \leq p \leq +\infty$, defines a tempered distribution by setting

$$<f, \phi> = \int_{\underline{R}^n} f \cdot \phi \, dx, \ \forall \phi \in S(\underline{R}^n).$$

Indeed, we have

$$|<f, \phi| \leq \|f\|_p \cdot \|\phi\|_q$$

$(p^{-1}+q^{-1} = 1)$, by Hölder's inequality. By observing that if a sequence (ϕ_j) converges to zero in S, it also converges to zero in L^q, we conclude that every $f \in L^p$ defines a continuous linear functional on S. Moreover, it can be seen that L^p can be identified with a vector subspace of S'.

3. Every polynomial $P(x)$ with constant coefficients defined a tempered distribution.

In fact, it suffices to show that every monomial x^α defines a tempered distribution by setting

$$<x^\alpha, \phi> = \int_{\underline{R}^n} x^\alpha \phi(x) \, dx, \ \forall \phi \in S.$$

4. We say that a continuous function $f(x)$ is *slowly increasing at infinity* if there exists an integer $k \geq 0$ such that $(1 + r^2)^{-k/2} f(x)$ is bounded in $\underline{R}^n$. Or, equivalently, if there is a polynomial $P(x)$ such that $|f(x)| \leq |P(x)|$, $x \in \underline{R}^n$.

Every continuous function f slowly increasing at infinity defines a tempered distribution. In fact, set

$$<f, \phi> = \int_{\underline{R}^n} f(x)\phi(x)\ dx, \forall \phi \in S$$

We have

$$|<f,\phi>| \le \int_{\underline{R}^n} |f(x)\phi(x)|\ dx = \int_{\underline{R}^n} |(1+r^2)^{-k/2} f(x)\cdot(1+r^2)^{k/2}\phi(x)|\ dx$$

$$\le C \int_{\underline{R}^n} |(1+r^2)^{k/2}\phi(x)|\ dx.$$

By observing that if a sequence (ϕ_j) converges to zero in S then, for every $k \ge 0$, $((1+r^2)^{k/2}\ \phi_j)$ converges to zero in L^1, we conclude that f defines an element of S'.

5. Any derivative (in the sense of distributions) of a continuous function f slowly increasing at infinity defines a tempered distribution. Let $T = \partial^\alpha f$ and define

$$<T,\phi> = (-1)^{|\alpha|} \int_{\underline{R}^n} f(x)\cdot\partial^\alpha\phi(x)\ dx, \forall \phi \in S$$

We have

$$|<T,\phi>| \le \int_{\underline{R}^n} |f(x)|\ |\partial^\alpha\phi(x)|\ dx$$

$$= \int_{\underline{R}^n} |(1+r^2)^{-k/2} f(x)(1+r^2)^{k/2}\partial^\alpha\phi(x)|\ dx \le C \int_{\underline{R}^n} |(1+r^2)^{k/2}\partial^\alpha\phi(x)|\ dx$$

Again, observe that if a sequence (ϕ_j) converges to zero in S then for every $k \ge 0$ and every $\alpha \in N^n$, the sequence $((1+r^2)^{k/2}\ \partial^\alpha\phi_j)$ converges to zero in L^1. Hence, $T = \partial^\alpha f$ defines a tempered distribution.

Conversely, we shall prove in Chapter 6 that *every tempered distribution is the derivative of a continuous function slowly increasing at infinity*. It is precisely this property that motivates the name *tempered* given to the distributions belonging to S'.

3. THE FOURIER TRANSFORM IN $S(\underline{R}^n)$

Definition 4.3. *The Fourier transform of a function* f $\in S(\underline{R}^n)$ *is defined by the integral*

$$\hat{f}(\xi) = \int_{\underline{R}^n} e^{-i\langle x,\xi\rangle} f(x)\, dx \tag{4.4}$$

where $\langle x, \xi\rangle = x_1\xi_1 + \cdots + x_n\xi_n$.

Since f $\in S$, the integral (4.4) is absolutely convergent. The Fourier transform of f will also be denoted by Ff.

Example. As an exercise, the reader should prove that the Fourier transform of the function

$$\exp -\left(\frac{|x|^2}{2}\right)$$

is equal to the function $(2\pi)^{n/2} \exp - \frac{|\xi|^2}{2}$.

Properties. 1. *Let* f $\in S(\underline{R}^n)$. *The Fourier transform of the product* ix_jf *is equal to the partial derivative* $-\partial\hat{f}/\partial\xi_j$.

In fact, by differentiating both sides of (4.4) with respect to ξ_j and by observing that we can move the derivative inside the integral sign, we get

$$\frac{\partial\hat{f}(\xi)}{\partial\xi_j} = \int_{\underline{R}^n} \frac{\partial}{\partial\xi_j}(e^{-i\langle x,\xi\rangle} f(x))\, dx$$

$$= -\int_{\underline{R}^n} e^{-i\langle x,\xi\rangle}(ix_jf(x))\, dx = \widehat{ix_jf}. \quad \text{Q.E.D.}$$

2. *The Fourier transform of the partial derivative* $\partial f/\partial x_j$ *is equal to the product* $i\xi_j\hat{f}$.

In fact, integrating by parts, we get

$$\widehat{\frac{\partial f}{\partial x_j}}(\xi) = \int_{\underline{R}^n} e^{-i\langle x,\xi\rangle} \frac{\partial f}{\partial x_j}(x)\, dx$$

$$= i\xi_j \int_{\underline{R}^n} e^{-i\langle x,\xi\rangle} f(x)\, dx = i\xi_j\hat{f}. \quad \text{Q.E.D.}$$

Observe that if we use the notation

$$D_j = \frac{1}{i}\partial_j,\ 1 \le j \le n,$$

we get from Properties 1 and 2

$$\widehat{x_j f} = -D_j \hat{f} \text{ and } \widehat{D_j f} = \xi_j \hat{f}.$$

More generally, if $\alpha = (\alpha_1, \cdots, \alpha_n)$ we have

$$\widehat{x^\alpha f} = (-D)^\alpha \hat{f} \text{ and } \widehat{D^\alpha f} = \xi^\alpha \hat{f}.$$

Also, if

$$P(D) = \sum a_p D^p$$

is a partial differential operator with constant coefficients we have

$$\widehat{P(D)f} = P(\xi)\cdot\hat{f}(\xi).$$

Theorem 4.3. *The Fourier transform defines a continuous linear map from* $S(\underline{R}^n)$ *into* $S(\underline{R}^n)$.

Proof. Properties 1 and 2 above imply that

$$\xi^\alpha D^p \hat{f}(\xi) = \int_{\underline{R}^n} e^{-i<x,\xi>} D^\alpha((-x)^p f(x))\, dx.$$

By Proposition 4.1, $D^\alpha((-x)^p f(x))$ belongs to $S(\underline{R}^n)$; hence by (4.3),

$$(1+r^2)^k |D^\alpha((-x)^p f(x))|$$

is uniformly bounded in $\underline{R}^n$ for all integer $k \geq 0$. If we choose k such that

$$\int_{\underline{R}^n} \frac{1}{(1+r^2)^k}\, dx = C < +\infty,$$

we get the following inequality:

$$|\xi^\alpha D^p \hat{f}(\xi)| \le \int_{\underline{R}^n} |D^\alpha(x^p f(x))|\, dx$$

$$= \int_{\underline{R}^n} \frac{1}{(1+r^2)^k} (1+r^2)^k |D^\alpha(x^p f(x))|\, dx$$

$$\le C \cdot \sup_{x \in \underline{R}^n} (1+r^2)^k |D^\alpha(x^p f(x))|$$

which proves that for every α and for every p, $\xi^\alpha D^p \hat{f}(\xi)$ is uniformly bounded in $\underline{R}^n$; hence $f \in S(\underline{R}^n)$. On the other hand, the same inequality shows, taking into account Theorem 4.1, that if (f_j) is a sequence converging to zero in $S(\underline{R}^n)$, then the sequence of Fourier transforms $(\hat{f}_j)$ converges to zero in $S(\underline{R}^n)$; therefore,

$$F\colon\ S(\underline{R}^n) \to S(\underline{R}^n)$$

is a continuous linear map. Q.E.D.

The Inverse Fourier Transform

If $g \in S(\underline{R}^n)$ the integral

$$(2\pi)^{-n} \int_{\underline{R}^n} e^{i\langle x,\xi\rangle}\, g(x)\, dx \tag{4.5}$$

is absolutely convergent and defines a function of the variables $(\xi_1,\cdots,\xi_n)$ which belongs to $S(\underline{R}^n)$. Denoting by

$$(F^{-1}g)(\xi)$$

the integral (4.5) we can prove, as in Theorem 4.3, that F^{-1} defines a continuous linear map from $S(\underline{R}^n)$ into $S(\underline{R}^n)$.

The following relations are easily checked:

$$\overline{Fg} = (2\pi)^n F^{-1}\bar{g} \text{ and } F\bar{g} = (2\pi)^n \overline{F^{-1}g}. \tag{4.6}$$

Theorem 4.4. *(Fourier's inversion formula). We have*

$$f = F^{-1}(Ff), \ \forall f \in S$$

Proof. From Fubini's theorem we derive the following relation

$$\int_{\mathbb{R}^n} \hat{f}(\xi)g(\xi)e^{i<x,\xi>} d\xi = \int_{\mathbb{R}^n} g(\xi)e^{i<x,\xi>}\left\{\int_{\mathbb{R}^n} e^{-i<y,\xi>} f(y)\, dy\right\} d\xi$$

$$= \int_{\mathbb{R}^n}\int_{\mathbb{R}^n} e^{i<x-y,\xi>} g(\xi)f(y)\, d\xi\, dy = \int_{\mathbb{R}^n} \hat{g}(y-x)f(y)\, dy,$$

for all $f,g \in S(\mathbb{R}^n)$.

Changing variables, we can write (4.7) as follows:

$$\int_{\mathbb{R}^n} \hat{f}(\xi)g(\xi)e^{i<x,\xi>} d\xi = \int_{\mathbb{R}^n} \hat{g}(y)f(x+y)\, dy. \tag{4.8}$$

Replacing $g(\xi)$ by $g(\varepsilon\xi)$, whose Fourier transform is equal to $\varepsilon^{-n}\hat{g}(y/\varepsilon)$, and changing variables we get

$$\int_{\mathbb{R}^n} \hat{f}(\xi)g(\varepsilon\xi)e^{i<x,\xi>} d\xi = \int_{\mathbb{R}^n} \hat{g}(y)f(x+\varepsilon y)\, dy.$$

Letting $\varepsilon \to 0$ in the last relation, we obtain

$$g(0)\int_{\mathbb{R}^n} \hat{f}(\xi)e^{i<x,\xi>} d\xi = f(x)\int_{\mathbb{R}^n} \hat{g}(y)\, dy. \tag{4.9}$$

Now choosing $g(x) = \exp(-|x|^2/2)$ and noticing that

$$\hat{g}(y) = (2\pi)^{n/2} \exp - \frac{|y|^2}{2}$$

and that (Gauss integral)

$$\int_{\underline{R}^n} \exp - \frac{|x|^2}{2}\, dx = (2\pi)^{n/2}$$

we get

$$(2\pi)^n f(x) = \int_{\underline{R}^n} e^{i\langle x,\xi\rangle} \hat{f}(\xi)\, d\xi. \quad \text{Q.E.D.} \tag{4.10}$$

Theorems 4.3 and 4.4 imply the following.

Corollary. The Fourier transform F *defines a topological isomorphism from* $S(\underline{R}^n)$ *onto* $S(\underline{R}^n)$.

Properties of the Fourier Transform

(I) *If* $f, g \in S(\underline{R}^n)$ *we have*

$$\int_{\underline{R}^n} \hat{f}\cdot g = \int_{\underline{R}^n} \hat{f}\cdot g. \tag{4.11}$$

Indeed, it suffices to set $x = 0$ in (4.8). Q.E.D.

(II) *Parseval's formula. If* $f, g \in S(\underline{R}^n)$ *we have*

$$\int_{\underline{R}^n} f\cdot\bar{g} = (2\pi)^{-n} \int_{\underline{R}^n} \hat{f}\cdot\bar{\hat{g}}.$$

Indeed, using (I), the relations (4.6), and Fourier's inversion formula we obtain

$$\int_{\underline{R}^n} Ff\cdot\overline{Fg} = \int_{\underline{R}^n} f\cdot F(\overline{Fg}) = (2\pi)^n \int_{\underline{R}^n} f\cdot\overline{F^{-1}(Fg)} = (2\pi)^n \int_{\underline{R}^n} f\cdot\bar{g}. \quad \text{Q.E.D.}$$

(III) *The Fourier transform of a convolution. If* f *and* g *belong to* $S(\underline{R}^n)$ *we have*

$$\widehat{f*g} = \hat{f}\cdot\hat{g}.$$

Proof. By Fubini's theorem we get

$$\widehat{f*g}(\xi) = \int_{\underline{R}^n} e^{-i<x,\xi>}(f*g)(x)\,dx$$

$$= \int_{\underline{R}^n} e^{-i<x,\xi>}\left\{\int_{\underline{R}^n} f(y)g(x-y)\,dy\right\}dx$$

$$= \int_{\underline{R}^n}\int_{\underline{R}^n} e^{-i<x-y,\xi>}e^{-i<y,\xi>}f(y)g(x-y)\,dx\,dy$$

$$= \int_{\underline{R}^n} e^{-i<y,\xi>}\left\{\int_{\underline{R}^n} e^{-i<x-y,\xi>}g(x-y)\,dx\right\}f(y)\,dy$$

$$= \hat{g}(\xi)\cdot\int_{\underline{R}^n} e^{-i<y,\xi>}f(y)\,dy = \hat{f}(\xi)\cdot\hat{g}(\xi). \quad \text{Q.E.D.}$$

(IV) *The Fourier transform of a product. If* f,g $\in \phi(\underline{R}^n)$ *we have*

$$\widehat{f\cdot g} = (2\pi)^{-n}\hat{f}*\hat{g}.$$

Proof. By using Fourier's inversion formula and Fubini's theorem, we get

$$\widehat{f\cdot g}(\xi) = \int_{\underline{R}^n} e^{-i<x,\xi>}f(x)\cdot g(x)\,dx$$

$$= (2\pi)^{-n}\int_{\underline{R}^n} e^{i<x,\xi>}g(x)\left\{\int_{\underline{R}^n} e^{i<x,\eta>}\hat{f}(\eta)\,d\eta\right\}dx$$

$$= (2\pi)^{-n} \int_{\underline{R}^n} \int_{\underline{R}^n} e^{-i<x,\, \xi-\eta>} g(x)\hat{f}(\eta)\, dx\, d\eta$$

$$= (2\pi)^{-n} \int_{\underline{R}^n} \left\{\int_{\underline{R}^n} e^{-i<x,\xi-\eta>} g(x)\, dx\right\} \hat{f}(\eta)\, d\eta$$

$$= (2\pi)^{-n} \int_{\underline{R}^n} \int_{\underline{R}^n} \hat{g}(\xi - \eta)\hat{f}(\eta) d\eta = (2\pi)^{-n} (\hat{f}*\hat{g})(\xi). \quad \text{Q.E.D.}$$

4. FOURIER TRANSFORM OF TEMPERED DISTRIBUTIONS

We have seen in the previous section that the Fourier transform F is an isomorphism of $S(\underline{R}^n)$ onto $S(\underline{R}^n)$. This allows us to define, by duality, the Fourier transform of tempered distributions.

Definition 4.4. *Let* $T \in S'(\underline{R}^n)$. *Its Fourier transform is the tempered distribution* FT *defined by*

$$<FT, \phi> = <T, F\phi> \; \forall \phi \in S(\underline{R}^n).$$

The linear map from $S'(\underline{R}^n)$ into $S'(\underline{R}^n)$ defined by (4.12) is indeed the *transpose* of the isomorphism $F: S(\underline{R}^n) \to S(\underline{R}^n)$. The reason we are using the same F to represent either the Fourier transform of elements of S or the Fourier transform of tempered distributions is that when

$$T = \psi \in S(\underline{R}^n)$$

then, by Property I, it follows that the definition in (4.12) coincides with that in (4.4).

Since $F: S(\underline{R}^n) \to S(\underline{R}^n)$ is continuous, it follows that $F: S'(\underline{R}^n) \to S'(\underline{R}^n)$ is also a continuous map in the sense of the strong topology of $S'(\underline{R}^n)$. As we did for Fourier transforms of functions of $S(\underline{R}^n)$, we shall often denote the Fourier transform FT of $T \in S'(\underline{R}^n)$ by $\hat{T}$.

In the same way, we define the inverse Fourier transform of tempered distributions by duality:

$$<F^{-1}T,\phi> = <T,F^{-1}\phi>, \forall\phi \in S(\underline{R}^n).$$

It is also clear that F^{-1} is a continuous linear map from $S'(\underline{R}^n)$ into $S'(\underline{R}^n)$ equipped with the strong topologies and that the Fourier inversion formula

$$T = F^{-1}(FT), \ \forall T \in S'(\underline{R}^n),$$

holds true. Therefore, F is an isomorphism from $S'(\underline{R}^n)$ onto $S'(\underline{R}^n)$.

In the classical theory of Fourier transforms it is shown that if $f \in L^2(\underline{R}^n)$ then

$$\hat{f}(\xi) = \int_{\underline{R}^n} e^{-i<x,\xi>} f(x)\, dx$$

exists, belongs to $L^2(\underline{R}^n)$, and the map

$$f \in L^2(\underline{R}^n) \to \hat{f} \in L^2(\underline{R}^n)$$

is an isomorphism [2]. On the other hand, $L^2(\underline{R}^n)$ is a subspace of $S'(\underline{R}^n)$ (Example 2 above), hence the Fourier transform of square integrable functions also exists in the sense of distributions. It follows from Property I, which holds true in the classical case, that these two notions are actually the same. Within the framework of tempered distributions we shall prove the following

Theorem 4.5. *Let* $f \in L^2(\underline{R}^n)$ *and let* $\hat{f}$ *be its Fourier transform in the sense of distributions. We have:*

1. $\hat{f} \in L^2(\underline{R}^n)$.

2. $\|\hat{f}\|_{L^2(\underline{R}^n)} = (2\pi)^{n/2}\|f\|_{L^2(\underline{R}^n)}$ (Parseval's formula).

Proof. for all $\phi \in S$, we have

$$|\langle\hat{f},\phi\rangle| = |\langle f,\hat{\phi}\rangle|$$

$$= \left|\int f\cdot\hat{\phi}\right| \le \|f\|_{L^2(\underline{R}^n)} \cdot \|\hat{\phi}\|_{L^2(\underline{R}^n)} \le (2\pi)^{n/2}\|f\|_{L^2(\underline{R}^n)} \cdot \|\phi\|_{L^2(\underline{R}^n)},$$

using Property II. Such inequality implies that $\hat{f} \in L^2(\underline{R}^n)$ and that

$$\|\hat{f}\|_{L^2(\underline{R}^n)} \le (2\pi)^{n/2}\|f\|_{L^2(\underline{R}^n)}, \quad \forall f \in L^2(\underline{R}^n). \tag{4.13}$$

In a similar way, we can show that

$$\|F^{-1}f\|_{L^2(\underline{R}^n)} \le (2\pi)^{-n/2}\|f\|_{L^2(\underline{R}^n)}, \quad \forall f \in L^2(\underline{R}^n). \tag{4.14}$$

From inequalities (4.13) and (4.14) we get

$$\|f\|_{L^2(\underline{R}^n)} = \|F^{-1}(Ff)\|_{L^2(\underline{R}^n)} \le (2\pi)^{-n/2}\|Ff\|_{L^2(\underline{R}^n)} \le \|f\|_{L^2(\underline{R}^n)};$$

hence

$$\|\hat{f}\|_{L^2(\underline{R}^n)} = (2\pi)^{n/2}\|f\|_{L^2(\underline{R}^n)}. \quad \text{Q.E.D.}$$

Later on we shall need the following classical result about Fourier transforms of integrable functions.

Proposition 4.2. *If* $f \in L^1(\underline{R}^n)$, then

$$\hat{f}(\xi) = \int_{\underline{R}^n} e^{-i\langle x,\xi\rangle} f(x)\, dx$$

is a well-defined continuous function of $\xi \in \underline{R}^n$.

Proof. Since the integral is absolutely convergent, $\hat{f}(\xi)$ is well defined. By estimating

$$\hat{f}(\xi) - \hat{f}(\xi_0) = \int_{\underline{R}^n} e^{-i<x,\xi_0>}(e^{-i<x,\xi-\xi_0>}-1)f(x)\,dx,$$

we get the inequality

$$|\hat{f}(\xi) - \hat{f}(\xi_0)| \leq \int_{\underline{R}^n} |e^{-i<x,\xi-\xi_0>} - 1|\,|f(x)|\,dx$$

$$\leq 2\int_{\underline{R}^n} |\sin \frac{<x,\xi-\xi_0>}{2}|\,|f(x)|\,dx$$

$$\leq 2\int_{|\xi|\leq A} |\sin \frac{<x,\xi-\xi_0>}{2}|\,|f(x)|\,dx + 2\int_{|\xi|>A} |f(x)|\,dx$$

which implies that $|\hat{f}(\xi) - \hat{f}(\xi_0)|$ can be made arbitrarily small provided that we choose $|\xi - \xi_0|$ sufficiently small. Q.E.D.

5. THE FOURIER TRANSFORM OF A DISTRIBUTION WITH COMPACT SUPPORT

Our aim in this section is to prove that the Fourier transform of every distribution with compact support is a C^∞ function in $\underline{R}^n$ that can be extended to the complex space $\underline{C}^n$ as an entire analytic function.

Let $\zeta = (\zeta_1 \cdots, \zeta_n)$ denote a variable element of $\underline{C}^n$, where $\zeta_j = \xi_j + i\eta_j$ $(1 \leq j \leq n)$, ξ_j, $\eta_j \in \underline{R}^n$, $i = \sqrt{-1}$. Let

$$\frac{\partial}{\partial\zeta_j} = \frac{1}{2}\left(\frac{\partial}{\partial\xi_j} - i\frac{\partial}{\partial\eta_j}\right), \quad \frac{\partial}{\partial\bar{\zeta}_j} = \frac{1}{2}\left(\frac{\partial}{\partial\xi_j} + i\frac{\partial}{\partial\eta_j}\right),$$

$$\frac{\partial}{\partial\zeta} = \left(\frac{\partial}{\partial\zeta_1}, \cdots, \frac{\partial}{\partial\zeta_n}\right), \qquad \frac{\partial}{\partial\bar{\zeta}} = \left(\frac{\partial}{\partial\bar{\zeta}_1}, \cdots, \frac{\partial}{\partial\bar{\zeta}_n}\right).$$

If $p \in \underline{N}^n$, we set $(\partial/\partial\zeta)^p = (\partial/\partial\zeta_1)^{p}\cdots(\partial/\partial\zeta_n)^{p_n}$.

Definition 4.5. *Let* U *be an open subset of* $\underline{C}^n$. *A function* f: U → $\underline{C}$ *is said to be holomorphic at a point* $\zeta_0 \in U$ *if it can be represented as a power series in* $\zeta - \zeta_0$,

$$f(\zeta) = \sum_p a_p(\zeta - \zeta_0)^p,$$

the series being convergent in some neighborhood of ζ_0. *The function* f *is said to be holomorphic in* U *if it is holomorphic at every point of* U. *An entire function is a holomorphic function in* $\underline{C}^n$.

An equivalent definition is the following one. A function f of class C^1 on U is holomrophic on U if it satisfies the Cauchy-Riemann equations

$$\frac{\partial f}{\partial \bar{\zeta}_j} = 0, \quad 1 \le j \le n.$$

(See [3, 15].)

The coefficient a_p in the series expansion above is given by

$$a_p = \frac{1}{p!}\left(\frac{\partial}{\partial\zeta}\right)^p f(\zeta_0).$$

They can also be obtained by *Cauchy's integral formulas*. Let $\zeta_0 \in U$ and consider the *polydisk*

$$D(r_1,\cdots,r_n) = \{\zeta \in \underline{C}^n\colon |\zeta_j - \zeta_j^0| \le r_j, 1 \le j \le n$$

which we assume to be contained in U. Then, for all $p \in \underline{N}^n$, we have

$$\frac{1}{p!}\left(\frac{\partial}{\partial\zeta}\right)^p f(\zeta_0) =$$

$$\frac{1}{(2\pi i)^n} \int_{|\zeta_1-\zeta_1^0|=r_1} \cdots \int_{|\zeta_n-\zeta_n^0|=r_n} \frac{f(\zeta)\, d\zeta_1 \cdots d\zeta_n}{(\zeta_1-\zeta_1^0)^{p_1+1} \cdots (\zeta_n-\zeta_n^0)^{p_n+1}} .$$

(See [3, 15].)

Definition 4.6. *Let* U *be an open set in* $\underline{C}^n$ *and let* E *be a topological vector space. A function* f: U → E *is said to be holomorphic in* U *if, for every* $\zeta_0 \in U$, f *can be represented as a power series in* $\zeta - \zeta_0$ *with coefficients in* E, *the series being convergent in some neighborhood of* ζ_0.

A holomorphic function with values in a topological vector space E is often called a *vector-valued holomorphic function.* When $U = \underline{C}^n$, f *is called a vector-valued entire function.*

If f: U → E is a vector-valued holomorphic function, then for every element e' belonging to the dual E' of E, the complex valued function $f_{e'}$ defined on U by

$$f_{e'}(\zeta) = \langle f(\zeta), e' \rangle$$

is holomorphic on U. Indeed, by applying the operator $\partial/\partial\bar{\zeta}_j$ to both sides of the last relation, by noticing that

$$\frac{\partial}{\partial\bar{\zeta}_j} \langle f(\zeta), e' \rangle = \langle \frac{\partial f}{\partial\bar{\zeta}_j} (\zeta), e' \rangle$$

[32, Theorem 27.1], and that f is, by assumption, holomorphic, we get $\partial f_{e'}/\partial\bar{\zeta}_j = 0$, $1 \le j \le n$, $\forall e' \in E'$.

A vector-valued function f: U → E such that, for every $e' \in E'$, the complex-valued function $f_{e'}$ is holomorphic in U, is said to be a *scalarly holomorphic function* in U. We just proved that every vector-valued holomorphic function is scalarly holomorphic. Let us just mention that, conversely, if E is a complete topological vector space every scalarly holomorphic function in U with values in E is a vector-valued holomorphic function. (See [13].)

Examples. 1. Let $x \in \underline{R}$, $\zeta = \xi + i\eta \in \underline{C}$. The function

$$\zeta \in \underline{C} \to e^{-ix\zeta} \in C^{\infty}(\underline{R})$$

is an *entire function of* ζ *with values in the locally convex space* $C^{\infty}(\underline{R})$. Indeed, for every $x \in \underline{R}$,

$$e^{-ix\zeta} = 1 + \frac{-ix\zeta}{1!} + \cdots + \frac{(-ix\zeta)^n}{n!} + \cdots$$

is obviously an entire function of ζ. Moreover, the series, as well as all its derivatives with respect to the variable x, converges uniformly on every compact subset of $\underline{R}$.

2. More generally, if $x = (x_1,\cdots,x_n) \in \underline{R}^n$, $\zeta = (\zeta_1,\cdots,\zeta_n) \in \underline{C}^n$, and $\langle x,\zeta\rangle = x_1\zeta_1 +\cdots+ x_n\zeta_n$, the function

$$\zeta \in \underline{C}^n \to e^{-i\langle x,\zeta\rangle} \in C^{\infty}(\underline{R}^n)$$

is a $C^{\infty}(\underline{R}^n)$*-valued entire function.*

Theorem 4.6. *If* $T \in E'(\underline{R}^n)$, *its Fourier transform is a* C^{∞} *function in* $\underline{R}^n$ *given by*

$$\hat{T}(\xi) = \langle T_x, e^{-i\langle x,\zeta\rangle}\rangle. \qquad (4.15)$$

Moreover, $\hat{T}(\xi)$ *can be extended to the complex space* $\underline{C}^n$ *as an entire analytic function given by*

$$\hat{T}(\zeta) = \langle T_x, e^{-i\langle x,\zeta\rangle}\rangle. \qquad (4.16)$$

Proof. 1. First we observe that the right-hand side of (4.15) is well defined because T is a distribution with compact support while $e^{-i\langle x,\xi\rangle}$ is a C^{∞} function of x. By Remark 2 following the proof of Lemma 3.2, it follows that the right-hand side of (4.15) is a C^{∞} function of $\xi \in \underline{R}^n$.

2. Let us prove that relation (4.15) holds. Let (α_j) be a regularizing sequence. By Theorem 3.3, we have $\alpha_j * T \to T$, as $j \to +\infty$, in $\mathcal{E}'$, hence in $\mathcal{S}'$; consequently $\alpha_j * T \to T$ in $\mathcal{S}'$. Let $\check{\alpha}_j(x) = \alpha_j(-x)$. By the remark following Lemma 3.4, we have

$$(\check{\alpha}_j * e^{-i<\cdot,\xi>})(x) \to e^{-i<x,\xi>} \text{ in } C^\infty(\underline{R}^n).$$

Since $T \in \mathcal{E}'$, it follows that

$$<T_x, (\check{\alpha}_j * e^{-i<\cdot,\xi>})(x)> \to <T_x, e^{-i<x,\xi>}>.$$

If we can prove that

$$\widehat{T*\alpha_j}(\xi) = <(T*\alpha_j)(x), e^{-i<x,\xi>}> = <T_x, (\check{\alpha}_j * e^{-i<\cdot,\xi>})(x)> \quad (4.17)$$

then (4.15) will follow by letting $j \to +\infty$. It then suffices to prove (4.17). As we already know, $T*\alpha_j$ is, for every j, a C^∞ function with compact support; hence, its Fourier transform is given by

$$\widehat{T*\alpha_j}(\xi) = \int_{\underline{R}^n} e^{-i<x,\xi>}(T*\alpha_j)(x)\, dx = <(T*\alpha_j)(x), e^{-i<x,\xi>}> .$$

On the other hand, by (3.14), $(T*\alpha_j)(x) = <T_y, \alpha_j(x - y)>$ and, by Theorem 3.1,

$$<(T*\alpha_j)(x), e^{-i<x,\xi>}> = <<T_y, \alpha_j(x-y)>, e^{-i<x,\xi>}>$$

$$= <T_y, <\alpha_j(x-y), e^{-i<x,\xi>}>> = <T_y, (\check{\alpha}_j * e^{-i<\cdot,\xi>})(y)>,$$

which implies (4.17).

3. As mentioned above, the function

$$\zeta \in \underline{C}^n \to e^{-i<x,\zeta>} \in C^\infty(\underline{R}^n)$$

is a vector-valued entire analytic function. On the other hand, T is an element of the dual of $C^\infty(\underline{R}^n)$; hence, the complex-valued function

$$\zeta \in \underline{C}^n \to \hat{T}(\zeta) = \langle T_x, e^{-i\langle x,\zeta\rangle}\rangle \in C$$

is an entire analytic function. Q.E.D.

Definition 4.7. *The Fourier-Laplace transform of a distribution* $T \in \mathcal{E}'(\underline{R}^n)$ *is the entire analytic function*

$$\hat{T}(\zeta) = \langle T_x, e^{-i\langle x,\zeta\rangle}\rangle. \tag{4.16}$$

6. THE PRODUCT OF A DISTRIBUTION BY A C^∞ FUNCTION

Definition 4.8. *Let* $T \in \mathcal{D}'(\underline{R}^n)$ *and let* $\alpha \in C^\infty(\underline{R}^n)$. *The product of* α *by* T *is the distribution* αT *defined by*

$$\langle \alpha T, \phi\rangle = \langle T, \alpha\phi\rangle \;\; \forall\, \phi \in C_c^\infty(\underline{R}^n). \tag{4.18}$$

It is easily seen that when T is defined by a locally integrable function then the product αT defined above coincides with the usual product of functions.

Let us remark that if it is not possible, in general, to define the product of two distributions S and T. Indeed, if S = f and T = g are locally integrable functions, f·g is not necessarily locally integrable; hence it does not define a distribution.

The following are easy consequences of Definition 4.8.

1. *The support of* αT *is contained in the intersection of the support of* α *and the support of* T.
2. *For every* $1 \le j \le n$ *we have*

$$\partial_j(\alpha T) = \partial_j\alpha \cdot T + \alpha \cdot \partial_j T.$$

Indeed,

$$\langle \partial_j(\alpha T),\phi\rangle = -\langle \alpha T,\partial_j\phi\rangle = -\langle T,\alpha\partial_j\phi\rangle$$

$$= -\langle T,\partial_j(\alpha\phi)\rangle + \langle T,\partial_j\alpha\phi\rangle$$

$$= \langle \partial_j T,\alpha\phi\rangle + \langle T,\partial_j\alpha\phi\rangle$$

$$= \langle \alpha\partial_j T + \partial_j\alpha T,\phi\rangle, \forall\ \phi \in C_c^\infty(\underline{R}^n).$$ Q.E.D.

As a consequence we get the Leibniz formula

$$\partial^p(\alpha T) = \sum_{r+s=p} \frac{p!}{r!s!}\, \partial^r\alpha\ \partial^s T.$$

3. *The bilinear map*

$$C^\infty(\underline{R}^n) \times \mathcal{D}'(\underline{R}^n) \ni (\alpha, T) \to \alpha T \in \mathcal{D}'(\underline{R}^n)$$

is separately continuous.

In fact, suppose that (T_j) converges strongly to zero in $\mathcal{D}'$. If B is a bounded set in $C^\infty(\underline{R}^n)$ then the set $\alpha B = \{\alpha\phi: \phi \in B\}$ is also bounded in $C^\infty(\underline{R}^n)$. Hence,

$$\langle \alpha T_j,\phi\rangle = \langle T_j,\alpha\phi\rangle \to 0$$

uniformly when $\phi \in B$; therefore, $\alpha T_j \to 0$ strongly in $\mathcal{D}'(\underline{R}^n)$. As an exercise, we leave to the reader the proof that αT is continuous with respect to the first variable. Similarly

4. *If* $\alpha \in C_c^\infty$ *(resp.* C^∞*) and* $T \in \mathcal{D}'$ *(resp.* $\mathcal{E}'$*), then* $\alpha T \in \mathcal{E}'$ *and the map*

$$(\alpha, T) \in C_c^\infty \times \mathcal{D}'\ (\text{resp. } C^\infty \times \mathcal{E}') \to \alpha T \in \mathcal{E}'$$

is separately continuous.

7. THE SPACE OF MULTIPLIERS OF $S'(\underline{R}^n)$

Let $T \in S'$ and suppose we ask for conditions on the C^∞ function α in order that $\alpha T \in S'$. Clearly this is always the case if $\alpha \in C_c^\infty(\underline{R}^n)$. However, if $\alpha(x) = \exp |x|^2$, it is not true that $\alpha T \in S'$, the reason being that an exponential function grows very fast at infinity. We make the following definition.

Definition 4.9. *We denote by* $O_M(\underline{R}^n)$ *the space of all* $\phi \in C^\infty(\underline{R}^n)$ *such that for every* $p \in \underline{N}^n$ *there is a polynomial* $P_p(x)$ *such that*

$$|\partial^p\phi(x)| \le |P_p(x)|, \ \forall x \in \underline{R}^n. \tag{4.19}$$

O_M *is said to be the space of* C^∞ *functions slowly increasing at infinity.*

Proposition 4.2. *Let* $\phi \in C^\infty(\underline{R}^n)$. *The following are equivalent conditions:*

1. *For every* $p = (p_1,\cdots,p_n) \in \underline{N}^n$ *there is a polynomial* $P_p(x)$ *such that*

$$|\partial^p\phi(x)| \le |P_p(x)|, \ \forall x \in \underline{R}^n.$$

2. *For all* $f \in S$ *the product* $\phi f \in S$

3. *For every n-tuple* $p = (p_1,\cdots,p_n)$ *and for every* $f \in S$ *the function* $\partial^p\phi\cdot f$ *is bounded in* $\underline{R}^n$.

Proof. (1) => (2) Let $p \in \underline{N}^n$. By the Leibniz formula we have

$$\partial^p(\phi\cdot f) = \sum_{q\le p} \frac{p!}{q!(p-q)!} \partial^p\phi\cdot\partial^{p-q}f.$$

Hence

$$(1+r^2)^k\partial^p(\phi\cdot f) = \sum_{q\le p} \frac{p!}{q!(p-q)!}\partial^q\phi\cdot(1+r^2)^k\partial^{p-q}f,$$

with k a positive real number. Since ϕ satisfies condition 1, there is an integer $N > 0$ such that

$$|\partial^q\phi(x)| \le C\ (1+r^2)^N\ \forall x \in \underline{R}^n,\ \forall q \le p.$$

Making the appropriate replacement, we get

$$\sup_{x\in\underline{R}^n}\ \sup_{|p|\le m} |(1+r^2)^k\partial^p(\phi\cdot f)(x)| \le C \sup_{x\in\underline{R}^n}\ \sup_{|s|\le m} |(1+r^2)^{k+N}\partial^s f(x)| \quad (4.20)$$

which implies, since $f \in S$, that $f\phi \in S$.

(2) => (3). We have

$$\partial_j\phi\cdot f = \partial_j(\phi\cdot f) - \phi\cdot\partial_j f,\ 1 \le j \le n.$$

Since $f \in S$ then, by condition 2, $\phi\cdot\partial_j f \in S$; hence $\partial_j\phi\cdot f \in S$. By the Leibniz formula and an induction argument we get that $\partial^p\phi\cdot f \in S$, $\forall p \in \underline{N}^n$; hence condition 3 holds.

(3) => (1). Suppose, by contradiction, that condition 1 does not hold. Then, for some $p \in \underline{N}^n$, $|\partial^p\phi|$ is not bounded by any polynomial. Hence, by induction, we can find a sequence (x_j) of elements of $\underline{R}^n$ with $|x_{j+1}| \ge |x_j| + 2$ and such that $|\partial^p\phi(x_j)| > (1 + |x_j|^2)^j$. Let $\gamma \in S$ be the function defined in Example 3 of Section 1. We clearly have

$$|\gamma(x_j)\partial^p\phi(x_j)| > \alpha(0) = 1, \forall\ j = 1,2,\cdots,$$

which contradicts condition 3. Q.E.D.

The Topology of O_M. This is the locally convex topology defined by the family of seminorms

$$\gamma_{f,p}(\phi) = \sup_{x\in\underline{R}^n} |f(x)\cdot\partial^p\phi(x)|$$

where $f \in S(\underline{R}^n)$ and $p \in \underline{N}^n$. It is clear that such topology does not have a countable basis. Also, it can be shown that $\mathcal{O}_M$ is a complete space.

A sequence (or filter) (ϕ_j) converges to zero in $\mathcal{O}_M$ if and only if for every $f \in S$ and for every $p \in \underline{N}^n$, $(f(x)\partial^p\phi_j(x))$ converges to zero uniformly on $\underline{R}^n$. Or, equivalently, for every $f \in S$, $(f\phi_j)$ converges to zero in S.

From Definition 4.9 and Proposition 4.2 it follows that a set B is bounded in $\mathcal{O}_M$ if and only if for all $p \in \underline{N}^n$ there is a polynomial $P_p(x)$ such that

$$|\partial^p\phi(x)| \leq P_p(x), \forall\, x \in \underline{R}^n, \forall\, \phi \in B.$$

Proposition 4.3. *The bilinear map*

$$\mathcal{O}_M \times S \ni (\phi, f) \to \phi f \in S$$

is separately continuous.

Proof. 1. Fix $\phi \in \mathcal{O}_M$. The inequality (4.20) implies that the linear map

$$S \ni f \to \phi f \in S$$

is continuous.

2. Fix $f \in S$ and let (ϕ_j) be functions of $\mathcal{O}_M$ which converge to zero in $\mathcal{O}_M$. Then, for every $g \in S$ and for every $p \in \underline{N}^n$,

$$\gamma_{g,p}(\phi) = \sup_{x\in\underline{R}^n} |g(x)\partial^p\phi_j(x)| \to 0.$$

Let k be an integer and q an n-tuple of nonnegative integers. By the Leibniz formula

$$\partial^q(f\cdot\phi_j) = \sum_{q'+q''=q} \frac{q!}{q'!q''!}\, \partial^{q'}f\cdot\partial^{q''}\phi_j;$$

hence

$$\sup_{\substack{x\in R^n\\ |q|\le m}} |(1+r^2)^k \partial^q (f\cdot\phi_j)| \le \sum C_{q'q''} \sup_{\substack{x\in R^n\\ |q'|\le m\\ |q''|\le m}} |(1+r^2)^k \partial^{q'} f\cdot \partial^{q''}\phi_j|.$$

Since $(1+r^2)^k\partial^{q'} f \in S$, every term in the right-hand side converges to zero as $\phi_j \to 0$ in O_M; therefore, $f\cdot\phi_j \to 0$ in $S(\underline{R}^n)$. Q.E.D.

Remark that if $\phi \in C^\infty$ is such that $\phi f \in S$ for all $f \in S$, then, by Proposition 4.2, $\phi \in O_M$.

Theorem 4.7. *We have the following inclusions*

$$S(\underline{R}^n) \to O_M(\underline{R}^n) \to S'(\underline{R}^n)$$

with continuous imbeddings. Moreover, S is dense in O_M.

Proof. Every $\phi \in O_M$ defines a tempered distribution by

$$\langle\phi,f\rangle = \int_{\underline{R}^n} \phi f\, dx, \ \forall f \in S$$

Also, it is quite clear that $S \subset O_M$. Let (β_j) be a sequence of elements of $C_c^\infty(\underline{R}^n)$ such that $\beta_j = 1$ on K_j, where $(K_j)_{j=1,2,\cdots}$ is an increasing sequence of compact subsets of $\underline{R}^n$ whose union is $\underline{R}^n$. It easily follows that $\beta_j\phi \to \phi$ in O_M for all $\phi \in O_M$; hence S is dense in O_M. We leave to the reader the proof that the imbeddings are continuous. Q.E.D.

The space O_M is then a normal space of distributions (Definition 2.9).

Definition 4.10 *If* $\phi \in O_M$ *and* $T \in S'$ *we define the product* ϕT *by*

$$\langle\phi T,f\rangle = \langle T,\phi f\rangle, \ \forall f \in S.$$

As a consequence of Proposition 4.3, the product ϕT belongs to S' and we also have the following theorem, whose proof is left to the reader.

Theorem 4.8. *The bilinear map*

$$\mathcal{O}_M \times S' \ni (\phi,T) \to \phi T \in S'$$

is separately continuous.

From Proposition 4.4 and the reflexivity and the space S it follows that $\mathcal{O}_M$ is the *space of all multipliers* of S', i.e., we have the following result.

Proposition 4.5. *If* $\phi \in C^\infty$ *is such that* $\phi T \in S'$ *for all* $T \in S'$, *then* $\phi \in \mathcal{O}_M$.

Proof. Our assumption implies that for every $f \in S$ the map

$$T \to \langle \phi T, f \rangle$$

is a continuous linear functional on S'. Since S is a reflexive space, there is then an element $g \in S$ such that

$$\langle \phi T, f \rangle = \langle T, g \rangle, \ \forall\, T \in S'.$$

In particular,

$$\langle \phi\alpha, f \rangle = \langle \alpha, g \rangle, \ \forall \alpha \in C_c^\infty,$$

which implies that

$$\langle \alpha, \phi f \rangle = \langle \alpha, g \rangle, \ \forall \alpha \in C_c^\infty;$$

hence, $\phi f = g \in S$ for all $f \in S$. By the remark following the proof of Proposition 4.3, we get that $\phi \in \mathcal{O}_M$. Q.E.D.

8. SOME RESULTS ON CONVOLUTIONS WITH TEMPERED DISTRIBUTIONS

In Section 3 of Chapter 3, we defined the convolution of a C^∞ function (resp. a C^∞ function with compact support) and a distribution with compact support (resp. a distribution). The same definition extends to the case of a C^∞ function rapidly decreasing at infinity and a tempered distribution, namely, if $\phi \in \mathcal{S}(\underline{R}^n)$ and $T \in \mathcal{S}'(\underline{R}^n)$ we define

$$(T*\phi)(x) = \langle T_y, \phi(x - y)\rangle .$$

With a proof similar to that of Theorem 3.2 we can show that $T*\phi \in C^\infty(\underline{R}^n)$ and that the bilinear map

$$\mathcal{S} \times \mathcal{S}' \ni (\phi, T) \to T*\phi \in C^\infty$$

is separately continuous.

In general, it is *not* true that $T*\phi \in \mathcal{S}$. In fact, let T be the function identically equal to one on $\underline{R}^n$. Then,

$$(T*\phi)(x) = \langle 1_y, \phi(x - y)\rangle = \int_{\underline{R}^n} \phi(y)\, dy = C,$$

a constant; hence $T*\phi \notin \mathcal{S}$. However, by using the structure of tempered distributions mentioned at the end of Section 2 we can prove the following.

Theorem 4.9. *If* $\phi \in \mathcal{S}$ *and* $T \in \mathcal{S}'$ *then* $T*\phi \in \mathcal{O}_M$.

Proof. If $T \in \mathcal{S}'$ then

$$T = \partial^p((1+r^2)^{k/2} f),$$

where f is a bounded continuous function on $\underline{R}^n$ (Chapter 6, corollary of Theorem 6.5). By definition, we have

$$(T*\phi)(x) = \langle T_y, \phi(x - y)\rangle = \langle \frac{\partial^p}{\partial y^p}((1 + |y|^2)^{k/2} f(y)), \phi(x - y)\rangle$$

$$= (-1)^{|p|} \int_{\underline{R}^n} (1 + |y|^2)^{k/2} f(y) \frac{\partial^p \phi}{\partial y^p}(x - y)\, dy.$$

Using the inequality

$$(1 + |y|^2)^{k/2} \le C(1 + |x|^2)^{k/2} (1 + |x - y|^2)^{k/2}$$

(Chapter 5, Lemma 5.2) we get

$$|T*\phi(x)| \le C(1 + |x|^2)^{k/2} \int_{\underline{R}^n} (1 + |x - y|^2)^{k/2} | \frac{\partial^p \phi}{\partial x^p}(x - y)|\, dy$$

$$\le C'(1 + |x|^2)^{k/2}.$$

Similarly, we get

$$|\partial^q(T*\phi)| \le C_q(1 + |x|^2)^{k/2}, \ \forall q \in \underline{N}^n;$$

hence, by Proposition 4.2, $T*\phi \in \mathcal{O}_M$. Q.E.D.

Theorem 4.10. *If* $S \in \mathcal{S}'$ *and* $T \in \mathcal{E}'$ *then* $S*T \in \mathcal{S}'$. *Moreover, the bilinear map*

$$\mathcal{S}' \times \mathcal{E}' \ni (S,T) \to S*T \in \mathcal{S}'$$

is separately continuous.

The proof is based on the following lemma.

Lemma 4.1. *If* $T \in \mathcal{E}'$ *and* $\phi \in \mathcal{S}$ *the function*

$$\psi(\xi) = \langle T_\eta, \phi(\xi + \eta)\rangle$$

belongs to S. *Furthermore, if* ϕ *converges to zero in* S, *then* ψ *converges to zero in* S.

Proof. Since by Theorem 2.22 every distribution T with compact support can be written as a finite sum of derivatives of continuous functions with compact support contained in an arbitrary neighborhood of the support of T, it suffices to prove the lemma when

$$T = \partial^q G,$$

where G is a continuous function in $\underline{R}^n$ such that its support is contained in an arbitrary neighborhood of the support of T.

Since $\psi(\xi)$ is a C^∞ function in $\underline{R}^n$ (see Remark 2 following the proof of Lemma 3.2), it suffices to show that $\psi(\xi)$ is rapidly decreasing at infinity. We have

$$\psi(\xi) = \langle T_\eta, \phi(\xi + \eta)\rangle = (-1)^{|q|} \int_{\underline{R}^n} G(\eta) \frac{\partial^q}{\partial \eta^q} (\xi + \eta)\, d\eta;$$

hence,

$$(1 + |\xi|^2)^{k/2} \frac{\partial^p \psi}{\partial \xi^p}(\xi) = (-1)^{|q|} \int_{\underline{R}^n} G(\eta)(1 + |\xi|^2)^{k/2} \frac{\partial^{p+q}}{\partial \xi^p \partial \eta^q} (\xi + \eta)\, d\eta.$$

By using the inequality

$$1 + |\xi|^2 \leq 2\cdot(1 + |\eta|^2)(1 + |\xi+\eta|^2),$$

we can estimate the last expression as follows:

$$\left|(1 + |\xi|^2)^{k/2} \frac{\partial^p \psi}{\partial \xi^p}(\xi)\right|$$

$$\leq 2 \cdot \int_{\underline{R}^n} (1 + |\eta|^2)^{k/2} |G(\eta)| (1 + |\xi + \eta|^2)^{k/2} \left|\frac{\partial^{p+q}}{\partial \xi^p \partial \eta^q} (\xi + \eta)\right| d\eta;$$

hence, taking into account that G is a continuous function with compact support, we get the inequality

$$\sup_{|p| \leq m} \sup_{\xi \in \underline{R}^n} \left|(1 + |\xi|^2)^{k/2} \frac{\partial^p \psi}{\partial \xi^p}(\xi)\right|$$

$$\leq C \cdot \sup_{|s| \leq \ell} \sup_{\zeta \in \underline{R}^n} \left|(1 + |\zeta|^2)^{k/2} \frac{\partial^s \phi}{\partial \zeta^s}(\zeta)\right|$$

which shows at the same time that $\psi \in S$ and that it depends continuously on ϕ in S. Q.E.D.

Remark. By recalling that

$$(T * \phi)(x) = \langle T_y, \phi(x - y) \rangle,$$

one can show, with a proof similar to that of the previous lemma, that *if* $T \in E'$ *and* $\phi \in S$ *then* $T * \phi \in S$.

Proof of Theorem 4.10. 1. Let $S \in S'$, $T \in E'$, and $\phi \in S$. As in the proof of Theorem 3.2 we can show that

$$\langle S_\xi, \phi(\xi + \eta) \rangle$$

is an infinitely differentiable function of η depending continuously on $\phi \in S$. Therefore,

$$\langle T_\eta, \langle S_\xi, \phi(\xi + \eta) \rangle \rangle \tag{4.21}$$

is well defined and depends continuously on $\phi \in \mathcal{S}$. By Lemma 4.1 the function

$$\psi(\xi) = \langle T_\eta, \phi(\xi + \eta)\rangle$$

belongs to $\mathcal{S}$ and depends continuously on $\phi \in \mathcal{S}$. Since $S \in \mathcal{S}'$

$$\langle S_\xi, \langle T_\eta, \phi(\xi + \eta)\rangle\rangle \tag{4.22}$$

is also well defined and it depends continuously on $\phi \in \mathcal{S}$. On the other hand, since (4.21) and (4.22) do coincide when $\phi \in C_c^\infty$ (see the definition of the convolution product) and C_c^∞ is dense in $\mathcal{S}$ (Theorem 4.2), they coincide everywhere in $\mathcal{S}$. We then get

$$\begin{aligned}\langle S*T,\phi\rangle &= \langle S_\xi \otimes T_\eta, \phi(\xi + \eta)\rangle \\ &= \langle S_\xi,\langle T_\eta, \phi(\xi + \eta)\rangle\rangle = \langle T_\eta, \langle S_\xi, \phi(\xi + \eta)\rangle\rangle\end{aligned}$$

for all $\phi \in \mathcal{S}$; therefore $S*T \in \mathcal{S}'$.

2. Fix $T \in \mathcal{E}'(\underline{R}^n)$ and suppose that the distributions S_j converge to zero strongly in $\mathcal{S}'(\underline{R}^n)$. We have for every $\phi \in \mathcal{S}$

$$\langle S_j*T, \phi\rangle = ((S_j*T)*\check{\phi})(0) = T*(S_j*\check{\phi})(0). \tag{4.23}$$

As mentioned above, the convolution product of a function on $\mathcal{S}$ and a distribution of $\mathcal{S}'$ is separately continuous; hence, $S_j*\check{\phi} \to 0$ in C^∞. Furthermore, it can be shown (Problem 14) that $S_j*\check{\phi}$ converges uniformly to zero whenever ϕ belongs to a bounded set in $\mathcal{S}$. Hence, it follows that $T*(S_j*\check{\phi}) \to 0$ in C^∞ uniformly whenever ϕ belongs to a bounded set of $\mathcal{S}$. This implies, taking into account (4.23), that $S_j*T \to 0$ strongly in $\mathcal{S}'$ as $S_j \to 0$ strongly in $\mathcal{S}'$.

Analogously, the reader can prove the continuity of $S*T$ with respect to T. Q.E.D.

Going back to Fourier transforms prove the following result.

Theorem 4.11. *If* $\phi \in S$ *and* $T \in S'$ *then*

$$\widehat{\phi * T} = \hat{\phi} \cdot \hat{T}. \tag{4.24}$$

Proof. By theorem 4.9, $\widehat{\phi * T} \in O_M$ and by Theorem 4.7, $\phi * T \in S'$; hence, the Fourier transform $\phi * T$ is well defined and it belongs to S'. On the other hand, $\hat{\phi} \in S \subset O_M$ and $\hat{T} \in S'$; hence by Theorem 4.8, $\hat{\phi}\hat{T} \in S'$. It then remains to be shown that both sides of (4.24) are equal.

For all $\psi \in S$ we have

$$\langle\widehat{\phi * T}, \psi\rangle = \langle\phi * T, \hat{\psi}\rangle = \langle\phi(\xi) \otimes T_\eta, \hat{\psi}(\xi + \eta)\rangle$$

$$= \langle T_\eta, \langle\phi(\xi), \hat{\psi}(\xi + \eta)\rangle\rangle = \langle T_\eta, \langle\phi(\xi - \eta), \hat{\psi}(\xi)\rangle\rangle,$$

after an obvious change of variable inside the integral $\langle\phi(\xi), \hat{\psi}(\xi+\eta)\rangle$. On the other hand,

$$\langle\phi(\xi - \eta), \hat{\psi}(\xi)\rangle = \int_{\underline{R}^n} \phi(\xi - \eta)\, \hat{\psi}(\xi)\, d\xi = (\check{\phi} * \hat{\psi})(\eta)$$

and it is easy to check that

$$\check{\phi} = F(F^{-1}\phi) = (2\pi)^{-n}\hat{\hat{\phi}},$$

where $\hat{\hat{\phi}} = F(F\phi)$. Making the appropriate replacement,

$$\langle\phi(\xi - \eta), \hat{\psi}(\xi)\rangle = (2\pi)^{-n}(\hat{\hat{\phi}} * \hat{\psi}) = \widehat{(\hat{\phi} \cdot \psi)}(\eta)$$

by Property IV of Section 3. Thus, we can write

$$\langle\widehat{\phi * T}, \psi\rangle = \langle\phi * T, \hat{\psi}\rangle = \langle T_\eta, (\check{\phi} * \hat{\psi})(\eta)\rangle$$

$$= \langle T_\eta, \widehat{(\hat{\phi} \cdot \psi)}(\eta)\rangle = \langle\hat{T}, \hat{\phi} \cdot \psi\rangle = \langle\hat{\phi}\hat{T}, \psi\rangle, \quad \text{Q.E.D.}$$

9. THE PALEY-WIENER-SCHWARTZ THEOREM

Theorem 4.12. 1. *The Fourier-Laplace transform of a distribution* T *with compact support in* $\underline{R}^n$ *is an entire function* $F(\zeta)$ *in* $\underline{C}^n$ *satisfying the following property:*

(P_1) *There are constants* C *and* A *and an integer* $N \geq 0$ *such that*

$$|F(\zeta)| \leq C(1+|\zeta|)^N e^{A|\text{Im}\ \zeta|}, \ \forall\zeta \in \underline{C}^n. \tag{4.25}$$

Conversely, every entire function in $\underline{C}^n$ *satisfying Property* (P_1) *is the Fourier-Laplace transform of a distribution belonging to* $E'(\underline{R}^n)$.

2. *The Fourier-Laplace transform of an infinitely differentiable function with compact support in* $\underline{R}^n$ *is an entire function* $F(\zeta)$ *in* $\underline{C}^n$ *satisfying the following property:*

(P_2) *There is a constant* $A > 0$ *such that for every integer* $N \geq 0$ *we can find a constant* C *such that*

$$|F(\zeta)| \leq C(1+|\zeta|)^{-N} e^{A|\text{Im}\ \zeta|}, \ \forall\zeta \in \underline{C}^n. \tag{4.26}$$

Conversely, every entire function in $\underline{C}^n$ *satisfying Property* (P_2) *is the Fourier-Laplace transform of a* C^∞ *function with compact support in* $\underline{R}^n$.

Proof. (i) Suppose that

$$F(\zeta) = \langle T_x, e^{-i\langle x,\zeta\rangle}\rangle$$

is the Fourier-Laplace transform of a distribution $T \in E'(\underline{R}^n)$. By Theorem 2.14, there is a constant $C > 0$, an integer $N \geq 0$, and a compact subset K of $\underline{R}^n$ such that

$$|\langle T, \phi\rangle| \leq C \cdot \sup_{\substack{|\alpha|\leq N \\ x\in K}} |D^\alpha\phi|, \forall\ \phi \in C^\infty(\underline{R}^n). \tag{4.27}$$

Let A be a positive real number such that the compact subset K is contained in the closed ball

$$\{x \in \underline{R}^n : |x| \leq A\}$$

and let

$$\phi_\zeta(x) = e^{-i<x,\zeta>}.$$

We have

$$|D^\alpha \phi_\zeta(x)| \leq |\zeta|^{|\alpha|} e^{<x,\eta>}$$

hence

$$\sup_{\substack{|\alpha|\leq N \\ |x|\leq A}} |D^\alpha \phi_\zeta(x)| \leq \sup_{\substack{|\alpha|\leq N \\ |x|\leq A}} |\zeta|^{|\alpha|} e^{<x,\eta>} \leq (1 + |\zeta|)^N \cdot e^{A|\eta|}. \quad (4.28)$$

The inequalities (4.27) and (4.28), together with the definition of $F(\zeta)$, imply the inequality (4.25).

(ii) Suppose that $F(\zeta)$ is now the Fourier-Laplace transform of a function $\phi \in C_c^\infty(\underline{R}^n)$. Let $A > 0$ be such that the support of ϕ is contained in the ball with center at the origin and radius A. For every n-tuple $\alpha = (\alpha_1, \cdots, \alpha_n)$ of nonnegative integers we can write

$$\zeta^\alpha \hat{\phi}(\zeta) = \int_{\underline{R}^n} e^{-i<x,\zeta>} D^\alpha \phi(x)\, dx.$$

From this relation we get immediately the inequality

$$|\zeta^\alpha \hat{\phi}(\zeta)| \leq c \cdot e^{A|\eta|}$$

which implies the inequality

$$(1 + |\zeta|^2)^{N/2} \, |\hat{\phi}(\zeta)| \le C \cdot e^{A|\eta|}, \tag{4.29}$$

for every integer $N \ge 0$. Finally, by using the inequality

$$0 < c_0 \le \frac{(1+|\zeta|^2)^{N/2}}{(1+|\zeta|)^N} \le C_0$$

we can see that (4.29) is equivalent to (4.26)

(iii) Conversely, suppose that Property (P_2) holds and define

$$f(x) = (2\pi)^{-n} \int_{\underline{R}^n} F(\xi) e^{i<x,\xi>} d\xi,$$

i.e., $f(x)$ is the inverse Fourier transform of $F(\xi)$, $\xi \in \underline{R}^n$. Because of (4.26) the last integral is absolutely convergent. Also, the same inequality (4.26) implies that the integral

$$D^{\alpha} f(x) = (2\pi)^{-n} \int_{\underline{R}^n} F(\xi) \xi^{\alpha} e^{i<x,\xi>} d\xi$$

is absolutely convergent. Therefore, $f(x)$ is an infinitely differentiable function in $\underline{R}^n$. Let us show that f has a compact support Let $\eta = (\eta_1, \cdots, \eta_n)$ be an arbitrarily fixed point in $\underline{R}^n$. Since, by assumption, $F(\zeta)$ is an entire function satisfying the inequality (4.26), then we can also write

$$f(x) = (2\pi)^{-n} \int_{\underline{R}^n} F(\xi + i\eta) e^{i<x,\, \xi+i\eta>} d\xi,$$

where the integral is an absolutely convergent one. Take $N = n + 1$. By using (4.26), we get

$$|f(x)| \leq (2\pi)^{-n}C.e^{-<x,\eta>+A|\eta|} \int_{\underline{R}^n} (1+|\xi|)^{-n-1}d\xi;$$

hence

$$|f(x)| \leq C'e^{A|\eta|-<x,\eta>}.$$

If in the last inequality we set $\eta = tx$, we get

$$|f(x)| \leq C'\cdot\exp[-t(|x|^2 - A|x|)].$$

Letting $t \to +\infty$, we see that this inequalty implies that $f(x)$ must be zero for all $x \in \underline{R}^n$ with norm greater than A. Consequently, the support of $f(x)$ is contained in the closed ball with center at the origin and radius A.

(iv) Finally, suppose that $F(\zeta)$ is an entire function in $\underline{C}^n$ satisfying Property (P_1). Then, from (4.25) it follows that $F(\xi)$, $\xi \in \underline{R}^n$, is a function slowly increasing at infinity; hence it defines a tempered distribution in $\underline{R}^n$ whose inverse Fourier transform T is, as we already know, an element of $S'(\underline{R}^n)$. We want to prove that T has a compact support in $\underline{R}^n$. For this, let (α_j), $j = 1,2,\cdots$, be a regularizing sequence in $C_c^\infty(\underline{R}^n)$ such that the support of α_j is contained in the ball with center at the origin and radius j^{-1}. We have, by Theorem 4.11,

$$\widehat{T*\alpha_j} = \hat{T}\cdot\hat{\alpha}_j.$$

By using (4.25), which, by assumption, holds for $\hat{T} = F(\zeta)$, and (4.26), which holds for every $\hat{\alpha}_j$, we get that every $\hat{T}\cdot\hat{\alpha}_j$ is an entire function in $\underline{C}^n$ satisfying an inequality of the type (4.26), where the constant A is replaced by the constant $A + j^{-1}$. It then follows that $T*\alpha_j$ is an infinitely differentiable function with compact support contained in the ball of center at the origin of radius $A + j^{-1}$. Therefore, the distribution T which is the limit of the sequence of functions $(T*\alpha_j)$ must have its support contained in the ball of center at the origin and radius A. Q.E.D.

Corollary 1. *If* $T \in E'(\underline{R}^n)$, *its Fourier transform is an infinitely differentiable function slowly increasing at infinity.*

Proof. The assertion is a trivial consequence of (4.25), letting $|\mathrm{Im}\ \zeta| = 0$. Q.E.D.

Corollary 2. *Let* $T \in S'(\underline{R}^n)$. *The following are equivalent conditions:*

(i) T *has support contained in the closed set:*

$$\left\{x \in \underline{R}^n : |x_j| \leq A_j,\ 1 \leq j \leq n\right\}.$$

(ii) *The Fourier-Laplace transform* $F(\zeta)$ *of* T *is an entire function in* $\underline{C}^n$ *such that for any* $\varepsilon > 0$ *there are a constant* C_ε *and an integer* $N \geq 0$ *such that*

$$|F(\zeta)| \leq C_\varepsilon (1 + |\xi|)^N \exp\ [(A_1 + \varepsilon)|\eta_1| + \cdots + (A_n + \varepsilon)\ |\eta_n|]. \tag{4.30}$$

The proof, which is a variation of the proof of Theorem 4.12, is left to the reader as an exercise. We give the following definition.

Definition 4.11. *An entire function* $f(\zeta)$ *in* $\underline{C}^n$ *is said to be of an exponential type* $(A_1,\cdots,A_n)$ *if for every* $\varepsilon > 0$ *there is a constant* $C_\varepsilon > 0$ *such that*

$$|f(\zeta)| \leq C_\varepsilon \exp\ [(A_1 + \varepsilon)|\zeta_1| + \cdots + (A_n + \varepsilon)|\zeta_n|\ ], \tag{4.31}$$

for $\zeta \in \underline{C}^n$.

The Fourier-Laplace transform of a distribution with compact support is then an *entire function of exponential type.*

10. A RESULT ON THE FOURIER TRANSFORM OF A CONVOLUTION OF TWO DISTRIBUTIONS

We have seen (Section 3, Property III) that the Fourier transform $\mathcal{F}$ transforms the convolution of two functions of $\mathcal{S}$ into the product of their Fourier transforms. Also, Theorem 4.11 shows that the Fourier transform of $\phi * T$, with $\phi \in \mathcal{S}$ and $T \in \mathcal{S}'$ equals the product $\hat{\phi} \cdot \hat{T}$. We are going to extend this result to the convolution product of a tempered distribution and a distribution with compact support with the following theorem.

Theorem 4.13. *If* $S \in \mathcal{S}'$ *and* $T \in \mathcal{E}'$ *then*

$$\widehat{T * S} = \hat{T} \cdot \hat{S}. \tag{4.32}$$

Proof. By Theorem 4.10, $T * S \in \mathcal{S}'$. Since $T \in \mathcal{E}'$, by the Paley-Wiener-Schwartz theorem, $\hat{T} \in \mathcal{O}_M$, hence, $\hat{T} \cdot \hat{S}$ is well defined and belongs to $\mathcal{S}'$ (Theorem 4.8). In order to show that both sides of (4.32) we shall use a regularizing sequence.

Let (α_j) be a sequence in $C_c^\infty(\underline{R}^n)$ converging to δ in $\mathcal{E}'(\underline{R}^n)$ as $j \to +\infty$. By Theorem 3.3, the sequence of functions

$$\phi_j = \alpha_j * T \in C_c^\infty$$

converges to T in $\mathcal{E}'$. By Theorem 4.10, $\phi_j * S \to T * S$ in $\mathcal{S}'$; hence, taking Fourier transforms,

$$\widehat{T * S} = \lim \widehat{\phi_j * S} = \lim \hat{\phi}_j \cdot \hat{S}, \tag{4.33}$$

the last relation being a consequence of Theorem 4.11.

On the other hand, since $\phi_j = \alpha_j * T \to T$ in $\mathcal{E}'$, it also converges in $\mathcal{S}'$, because the imbedding of $\mathcal{E}'$ into $\mathcal{S}'$ is a continuous one (Theorem 4.2). Hence, by Fourier transform $\hat{\phi}_j \to \hat{T}$ in $\mathcal{S}'$. But the functions $\hat{\phi}_j$ and $\hat{T}$ do belong to $\mathcal{O}_M$ and $\hat{\phi}_j \to \hat{T}$ in $\mathcal{O}_M$. In fact, it suffices to show (see Problem 16) that for all $\hat{f} \in \mathcal{T}$, $\hat{\phi}_j \cdot \hat{f} \to \hat{T} \cdot \hat{f}$ in

$\mathcal{S}$ or, equivalently, by taking Fourier transforms, that $\phi_j * f \to T*f$ in $\mathcal{S}$, which is a trivial consequence of the remark following Lemma 4.1. Therefore, by Theorem 4.8, we have

$$\hat{T}\cdot\hat{S} = \lim \hat{\phi}_j \cdot \hat{S}, \tag{4.34}$$

the limit being taken in $\mathcal{S}'$. The relations (4.33) and (4.34) imply (4.32). Q.E.D.

Remark. The result of Theorem 4.13 is not the best one in that direction since it is possible to prove, without using the Paley-Wiener-Schwartz theorem, a more general theorem. In Chapter 6, we shall define the space $\mathcal{O}'_c$ which plays with respect to the convolution with tempered distributions the same role that the space $\mathcal{O}_M$ plays with respect to the product with tempered distributions. We shall see that $\mathcal{F}$ is a one-to-one map from $\mathcal{O}_M$ onto $\mathcal{O}'_c$ and vice versa and that $\mathcal{F}$ transforms the convolution of a distribution in $\mathcal{O}'_c$ with a tempered distribution into a product of an element of $\mathcal{O}_M$ by a tempered distribution.

PROBLEMS

1. Prove the equivalence of the conditions (4.1), (4.2), and (4.3).

2. Prove that the space $\mathcal{S}$ is complete.

3. Prove Theorem 4.1.

4. Show that B is a bounded subset in $\mathcal{S}$ if and only if, for all polynomials P(x) with constant coefficients and for all partial differential operators $Q(\partial)$ with constant coefficients there is a constant $C_{P,Q}$ such that

$$|P(x)Q(\partial)\phi(x)| \leq C_{P,Q}$$

for all $x \in \underline{R}^n$ and for all $\phi \in B$.

5. Prove that the imbeddings $\mathcal{E}'(\underline{R}^n) \to \mathcal{S}'(\underline{R}^n) \to \mathcal{D}'(\underline{R}^n)$ are continuous in the sense of the strong topologies.

6. Prove that: (i) if a sequence (ϕ_j) converges to zero in $\mathcal{S}(\underline{R}^n)$, it converges to zero in $L^q(\underline{R}^n)$, $1 \leq q \leq +\infty$; (ii) if (ϕ_j) converges to zero in $\mathcal{S}$, then for all $k \in N$ and all $\alpha \in \underline{N}^n$ the sequence $((1+r^2)^{k/2}\, \partial^\alpha \phi_j)$ converges to zero in L^q, $1 \leq q \leq +\infty$.

7. Prove that the Fourier transform of the function exp $(-|x|^2/2)$ is equal to the function $(2\pi)^{n/2}$ exp $(-|\xi|^2/2)$.

8. Show that $\mathcal{F}$: $\mathcal{S}' \to \mathcal{S}'$ is continuous in the sense of the strong topologies.

9. Prove the following formulas:

$$\mathcal{F}\delta = 1, \qquad \mathcal{F}1 = (2\pi)^n \delta;$$

$$\mathcal{F}(D_j \delta) = \xi_j, \ \mathcal{F}(x_j) = (2\pi)^n (-D_j)\delta.$$

10. If $T = f$ is a locally integrable function and if $\alpha \in C^\infty$, show that the product αT according to Definition 4.8 coincides with the usual product of functions.

11. Complete the proof of Property 3 of Section 6. Prove Property 4 of Section 6.

12. Prove that the imbeddings $\mathcal{S} \to \mathcal{O}_M \to \mathcal{S}'$ are continuous.

13. Let $T \in \mathcal{S}'(\underline{R}^n)$. Show that $\widehat{D^\alpha T} = \xi^\alpha \hat{T}$ and $\widehat{x^\alpha T} = (-D)^\alpha \hat{T}$.

14. If a sequence of tempered distributions (S_j) converges strongly to zero, then for all $\phi \in \mathcal{S}$, $(S_j * \phi)$ converges to zero in C^∞. Moreover, it converges uniformly to zero whenever ϕ belongs to a bounded subset of $\mathcal{S}$.

15. Complete the proof of Theorem 4.10 by proving the continuity of $S*T$ with respect to T.

16. Complete the proof of Theorem 4.13 by showing that $\hat{\phi}_j \to \hat{T}$ in $\mathcal{O}_M$.

17. Show that $\mathcal{F}^{-1}(f*g) = (2\pi)^n (\mathcal{F}^{-1}f)(\mathcal{F}^{-1}g)$ for all $f, g \in \mathcal{S}$.

18. If $\phi \in \mathcal{S}$ and $T \in \mathcal{S}'$, prove that $\mathcal{F}^{-1}(\phi * T) = (2\pi)^n (\mathcal{F}^{-1}\phi)(\mathcal{F}^{-1}T)$.

19. If $\phi \in S$ and $T \in S'$, prove that $\widehat{\phi T} = (2\pi)^{-n}\hat{\phi} * \hat{T}$.

20. Show that $\mathcal{O}_M$ is the space of all functions $\phi \in C^\infty(\underline{R}^n)$ such that, for all $p \in \underline{N}^n$, there is an integer $k \in \underline{Z}$ such that $(1 + |x|^2)^k \partial^p f(x)$ is bounded in $\underline{R}^n$.

21. If $\phi \in \mathcal{O}_M$ show that $\partial^p \phi \in \mathcal{O}_M$ for all $p \in \underline{N}^n$.

22. If $\phi \in C^\infty$ is such that $\phi f \in S$ for all $f \in S$, prove that $\phi \in \mathcal{O}_M$.

23. As a consequence of the Paley-Wiener-Schwartz theorem, prove that if $T \in E'(\underline{R}^n)$ and $\langle T,P \rangle = 0$ for all polynomial P with constant coefficients in $\underline{R}^n$, then $T \equiv 0$. (*Hint:* $D^p\hat{T}(0) = 0$, $\forall p \in \underline{N}^n$.). Hence, the space P of all polynomials with constant coefficients in $\underline{R}^n$ is dense in $C^\infty(\underline{R}^n)$.

Chapter 5

SOBOLEV SPACES

1. THE DEFINITION OF SOBOLEV SPACES

Let Ω be an open subset of $\underline{R}^n$, let m be a nonnegative integer, and let $1 \leq p \leq +\infty$.

Definition 5.1. *We denote* $H^{m,p}(\Omega)$ *the space of all distributions* $f \in \mathcal{D}'(\Omega)$ *such that*

$$D^\alpha f \in L^p(\Omega), \ \forall \ |\alpha| \leq m,$$

equipped with the norm

$$\|f\|_{m,p} = \left\{ \sum_{|\alpha| \leq m} \int_\Omega |D^\alpha f(x)|^p dx \right\}^{1/p}. \tag{5.1}$$

When m = 0, $H^{0,p}(\Omega) = L^p(\Omega)$. The following are easy consequences of Definition 5.1:

1. If $m \geq \ell$, $H^{m,p}(\Omega) \to H^{\ell,p}(\Omega)$ with continuous imbedding.

2. The topology defined on $H^{m,p}(\Omega)$ by the norm (5.1) is the coarsest one for which the maps

$$D^\alpha: H^{m,p}(\Omega) \to L^p(\Omega), \ \forall |\alpha| \leq m,$$

are continuous.

3. If p = 2, the norm (5.1) is induced by the scalar product

$$(f, g)_{m,2} = \sum_{|\alpha|\leq m} \int_\Omega D^\alpha f(x) \cdot \overline{D^\alpha g(x)}\, dx. \tag{5.2}$$

For simplicity, when dealing with the case $p = 2$, we shall denote by $(f, g)_m$ the scalar product (5.2) and by $\|f\|_m$ its corresponding norm. Analogously, we shall denote by $H^m(\Omega)$ the Sobolev space $H^{m,2}(\Omega)$.

Theorem 5.1. $H^{m,p}(\Omega)$ *is a Banach space.*

Proof. It suffices to show that $H^{m,p}(\Omega)$ is complete. Let $(f_j)_{j\in N}$ be a Cauchy sequence in $H^{m,p}(\Omega)$. For every $\alpha = (\alpha_1,\cdots,\alpha_n)$ with $|\alpha| \leq m$, $(D^\alpha f_j)_{j\in N}$ is a Cauchy sequence in $L^p(\Omega)$. Since the last space is a complete one,

$$D^\alpha f_j \to g_\alpha \text{ in } L^p(\Omega),\ \forall|\alpha| \leq m.$$

In particular, $f_j \to g_0$ in $L^p(\Omega)$; hence $f_j \to g_0$ in $\mathcal{D}'(\Omega)$. On the other hand, D^α is, for every α, a continuous linear operator from $\mathcal{D}'(\Omega)$ into $\mathcal{D}'(\Omega)$ (see Chapter 2, Section 5, Property 4). Hence

$$D^\alpha f_j \to D^\alpha g_0,\quad \forall|\alpha| \leq m.$$

By the uniqueness of the limit we get $g_\alpha = D^\alpha g_0$; therefore, $g_0 \in H^{m,p}(\Omega)$ and $g_j \to g_0$ in $H^{m,p}(\Omega)$. Q.E.D.

Definition 5.2. *We denote by* $H_0^{m,p}(\Omega)$ *the closure of* $C_c^\infty(\Omega)$ *in* $H^{m,p}(\Omega)$.

Being a closed subspace of a Banach space, $H_0^{m,p}(\Omega)$ is a Banach space (resp. Hilbert space when $p = 2$).

Observe that when $m = 0$ and $p < +\infty$, $H_0^{0,p}(\Omega) = L^p(\Omega)$, as is easy to see. However, in general, $H_0^{m,p}(\Omega)$ is a proper subspace of $H^{m,p}(\Omega)$.

Since, by definition,

$$C_c^\infty(\Omega) \subset H_0^{m,p}(\Omega) \subset \mathcal{D}'(\Omega)$$

with continuous imbeddings and $C_c^\infty(\Omega)$ dense in $H_0^{m,p}(\Omega)$, then $H_0^{m,p}(\Omega)$ is a normal space of distributions. Its dual, $(H_0^{m,p}(\Omega))'$, is a subspace of $\mathcal{D}'(\Omega)$ whose structure in the cases $1 \le p < +\infty$ is described by the following theorem.

Theorem 5.2. *The dual* $(H_0^{m,p}(\Omega))'$ *coincides with the space of all distributions* $T \in \mathcal{D}'(\Omega)$ *such that*

$$T = \sum_{|\alpha| \le m} \partial^\alpha f_\alpha$$

with $f_\alpha \in L^{p'}(\Omega)$, $(1/p) + (1/p') = 1$.

Proof. If T is of the form (5.3), define for all $\phi \in C_c^\infty(\Omega)$ the following linear functional:

$$\langle T, \phi\rangle = \sum_{|\alpha| \le m} (-1)^{|\alpha|} \langle f_\alpha, \partial^\alpha \phi\rangle = \sum_{|\alpha| \le m} (-1)^{|\alpha|} \int_\Omega f_\alpha \frac{\partial^\alpha \phi}{\partial x^\alpha}\, dx.$$

In view of Holder's inequality, we get

$$|\langle T, \phi\rangle| \le \sum_{|\alpha| \le m} \|f_\alpha\|_{L^{p'}} \|\partial^\alpha \phi\|_{L^p} \le C\|\phi\|_{m,p}$$

which shows, since $C_c^\infty(\Omega)$ is dense in $H_0^{m,p}(\Omega)$, that T defines a continuous linear functional on $H_0^{m,p}(\Omega)$.

Conversely, suppose that $T \in (H_0^{m,p}(\Omega)'$. Let N be the number of n-tuples $\alpha = (\alpha_1, \cdots, \alpha_n)$ such that $|\alpha| \le m$ and denote by $(L^p(\Omega))^N = L^p(\Omega)\times\cdots\times L^p(\Omega)$ (N times) the product of N copies of $L^p(\Omega)$ equipped with the product topology. Define the map

$$J\colon H^{m,p}(\Omega) \ni \phi \to (\partial^\alpha \phi)_{|\alpha| \le m} \in (L^p(\Omega))^N.$$

It is obviously a canonical isometry from $H^{m,p}(\Omega)$ into $(L^p(\Omega))^N$. In particular, $H^{m,p}(\Omega)$ can be identified, by J, with a subspace E of $(L^p(\Omega))^N$. Define on $E = J(H_0^{m,p}(\Omega))$ the following linear functional:

$$F((\partial^\alpha \phi)_{|\alpha|\leq m}) = \langle T, \phi \rangle, \forall \phi \in H_0^{m,p}(\Omega).$$

Since T is continuous on $H_0^{m,p}(\Omega)$, F is also continuous on E. By the Hahn-Banach theorem, we can extend F to a continuous linear functional on $(L^p(\Omega))^N$. But the dual of $(L^p(\Omega))^N$ can be identified with the product $(L^{p'}(\Omega))^N$. Hence, there are functions $g_\alpha \in L^{p'}(\Omega)$, $|\alpha| \leq m$, such that

$$\langle T, \phi \rangle = F((\partial^\alpha \phi)_{|\alpha|\leq m}) = \sum_{|\alpha|\leq m} \int_\Omega g_\alpha \frac{\partial^\alpha \phi}{\partial x^\alpha} dx, \ \forall \phi \in H_0^{m,p}(\Omega),$$

in particular, for all $\phi \in C_c^\infty(\Omega)$. Taking into account the definition of derivative in the sense of distributions we derive from the last relation that

$$\langle T, \phi \rangle = \sum_{|\alpha|\leq m} \partial^\alpha f_\alpha, \phi \ , \forall \phi \in C_c^\infty(\Omega),$$

where $f_\alpha = (-1)^{|\alpha|} g_\alpha$. Q.E.D.

We shall denote by $H^{-m,p'}(\Omega)$ the dual of $H_0^{m,p}(\Omega)$.

When $\Omega = \underline{R}^n$ and $1 \leq p < +\infty$, $H_0^{m,p}(\underline{R}^n) = H^{m,p}(\underline{R}^n)$. Indeed, this result is a consequence of the following theorem.

Theorem 5.3. *Let* $m \geq 0$ *and* $1 \leq p < +\infty$. *Then* $C_c^\infty(\underline{R}^n)$ *is a dense subspace of* $H^{m,p}(\underline{R}^n)$.

Proof. 1. Every element of $H^{m,p}(\underline{R}^n)$ can be approximated, in the sense of norm (5.1), by a sequence of elements of $H^{m,p}(\underline{R}^n)$ with compact support. In fact, if $f \in H^{m,p}(\underline{R}^n)$, let $(\beta_k)_{k=1,2,\cdots}$ be a sequence of functions belonging to C_c^∞ such that β_k is equal to one on the ball $B(0, k)$ and equal to zero outside the ball $B(0, k+1)$

and set $f_k = \beta_k f$, $k = 1,2,\cdots$. It is clear that $f_k \in L^p$ and that $f_k \to f$ in L^p. For all $1 \le j \le n$, we have

$$D_j f_k = \beta_k \cdot D_j f + D_j \beta_k \cdot f.$$

Since $D_j f \in L^p$, then $\beta_k D_j f \to D_j f$ in L^p. On the other hand, since $f \in L^p$ it follows from the definition of β_k that $D_j \beta_k \cdot f \in L^p$ and that $D_j \beta_k \cdot f \to 0$ in L^p as $k \to +\infty$. By induction, one can prove that $D^\alpha f_k \to D^\alpha f$ in L^p as $k \to +\infty$.

2. Let f be an element of $H^{m,p}(\underline{R}^n)$ with compact support and let $(\alpha_j)_{j=1,2,\cdots}$ be a sequence of functions of $C_c^\infty(\underline{R}^n)$ converging to δ in $\mathcal{D}'(R^n)$. By Theorem 1.1, the function $\phi_j = \alpha_j * f$ belongs to $C_c^\infty(\underline{R}^n)$ for all $j = 1,2,\cdots$. On the other hand, if $|\alpha| \le m$ we have, by formula (3.13),

$$D^\alpha \phi_j = \alpha_j * D^\alpha f.$$

Since $D^\alpha f \in L^p, \forall\ |\alpha| \le m$, by Theorem 1.1, part 4, it follows that

$$D^\alpha \phi_j \to D^\alpha f \text{ in } L^p \text{ as } j \to +\infty, \forall\ |\alpha| \le m.$$

Therefore, $\phi_j \to f$ in $H^{m,p}(\underline{R}^n)$. Q.E.D.

From now on, we shall concentrate our attention on the important case where $\Omega = \underline{R}^n$ and $p = 2$. In this case, by using the Fourier transform, it is possible to describe in a simple and elegant way the structure of Sobolev spaces as well as to get their most remarkable properties.

2. THE SOBOLEV SPACES $H^s(\underline{R}^n)$

Going back to Definition 1.1, we have that $f \in H^m(\underline{R}^n)$ if and only if $D^\alpha f \in L^2(\underline{R}^n)$. $\forall\ |\alpha| \le m$. By Theorem 4.5 and the properties of the Fourier transform the last condition is equivalent to

$$\xi^\alpha \hat{f}(\xi) \in L^2(\underline{R}^n), \forall\ |\alpha| \le m$$

which in turn is equivalent to the following one: *For every polynomial* $P(\xi)$ *with constant coefficients and degree* $\leq m$, $P(\xi)\hat{f} \in L^2(\underline{R}^n)$. Finally, this last condition is equivalent to

$$(1 + |\xi|^2)^{m/2}\, \hat{f}(\xi) \in L^2(\underline{R}^n).$$

These results motivate the following definition.

Definition 5.3. *Let* s *be a real number. We denote by* $H^s(\underline{R}^n)$ *the space of tempered distributions* $f \in S'(\underline{R}^n)$ *such that*

$$(1 + |\xi|^2)^{s/2}\, \hat{f}(\xi) \in L^2(\underline{R}^n)$$

equipped with the scalar product

$$(f, g)_s = \int_{\underline{R}^n} (1 + |\xi|^2)^{s/2}\, \hat{f}(\xi)\overline{\hat{g}(\xi)}\, d\xi \tag{5.4}$$

which induces the s*-norm*

$$\|f\|_s = \left(\int_{\underline{R}^n} (1 + |\xi|^2)^s |\hat{f}(\xi)|^2 d\xi \right)^{1/2}$$

It follows immediately that: (1) if $s \geq 0$, $H^s(\underline{R}^n) \subset L^2(\underline{R}^n)$; (2) if $s = m$, a nonnegative integer, then Definition 5.3 coincides with Definition 5.1 and the norms (5.1) and (5.5) are equivalent.

Theorem 5.4. *For every real number* s, $H^s(\underline{R}^n)$ *is a Hilbert space.*

Proof. Let (f_j) be a Cauchy sequence in $H^s(\underline{R}^n)$. By definition,

$$((1 + |\xi|^2)^{s/2}\, \hat{f}_j(\xi))$$

is a Cauchy sequence in $L^2(\underline{R}^n)$. Since the last space is complete, the sequence $((1 + |\xi|^2)^{s/2} \hat{f}_j(\xi))$ converges to $\hat{g}$ in $L^2(\underline{R}^n)$, hence in $S'(\underline{R}^n)$ (Chapter 4, Section 2, Example 2). Set

$$\hat{f}(\xi) = (1 + |\xi^2|)^{-s/2} \hat{g}(\xi).$$

It is clear that for any $s \gtreqless 0$, $\hat{f} \in S'(\underline{R}^n)$ and the sequence (f_j) converges to f in $H^s(\underline{R}^n)$. Q.E.D.

Theorem 5.5 *If* s *and* t *are real numbers such that* $s \geq t$, *we have the following inclusions*

$$S(\underline{R}^n) \subset H^s(\underline{R}^n) \subset H^t(\underline{R}^n) \subset S'(\underline{R}^n)$$
$$(s \geq t)$$

with continuous imbeddings. Moreover, S *is dense in* H^s *for all* s.

Proof. The imbedding

$$H^s \to H^t \quad (s \geq t)$$

follows immediately from Definition 5.3

Let $f \in H^s(\underline{R}^n)$. Since S is dense in L^2, given $\varepsilon > 0$, we can find $\phi \in S$ such that

$$\|(1 + |\xi|^2)^{s_2} \hat{f}(\xi) - \hat{\phi}(\xi)\|_{L^2} < \varepsilon.$$

On the other hand,

$$\hat{\psi}(\xi) = (1 + |\xi|^2)^{-s/2}\hat{\phi}(\xi)$$

belongs to S for all $s \gtreqless 0$. Making the appropriate replacement, we get

$$\|(1 + |\xi|^2)^{s/2} (\hat{f}(\xi) - \hat{\psi}(\xi))\|_{L^2} < \varepsilon,$$

i.e.,

$$\|f - \psi\|_s < \varepsilon;$$

therefore, $\mathcal{S}$ is dense H^s.

As an exercise, we leave to the reader the proof that the imbeddings

$$\mathcal{S} \to H^s \text{ and } H^s \to \mathcal{S}'$$

are continuous. Q.E.D.

As a consequence of Theorem 5.5 we get, in view of Theorem 4.2, that $C_c^\infty(\underline{R}^n)$ is dense in $H^s(\underline{R}^n)$ for all real s. Therefore, for every real s, $H^s(\underline{R}^n)$ can be also defined as a *completion of* $C_c^\infty(\underline{R}^n)$ *with respect to the norm* (5.5).

Theorem 5.6. *The strong dual of* $H^s(\underline{R}^n)$ *can be identified, in the algebraic and topological senses, with* $H^{-s}(\underline{R}^n)$.

Proof. First of all we observe that since

$$\mathcal{S} \subset H^s \subset \mathcal{S}'$$

with $\mathcal{S}$ dense in H^s for all s, then the dual $(H^s)'$ of H^s is a subspace of $\mathcal{S}'$.

If $f \in (H^s)'$, there is a constant $C > 0$ such that

$$|\langle f, \phi\rangle| \le C$$

for all $\phi \in \mathcal{S}$ such that $\|\phi\|_s \le 1$. Or equivalently, by Fourier transform,

$$|\langle\hat{f}, \hat{\phi}\rangle| \leq C_1, \quad \forall \ \|\phi\|_s \leq 1.$$

Set

$$\hat{\psi}(\xi) = (1 + |\xi|^2)^{s/2}\hat{\phi}(\xi).$$

Then, $\phi \in S$ with $\|\phi\|_s \leq 1$ implies that $\psi \in S$ with $\|\psi\|_{L^2} \leq 1$ and vice versa. By replacing in the last inequality we get

$$\left|\left\langle\hat{f}, \frac{\hat{\psi}}{(1 + |\xi|^2)^{s/2}}\right\rangle\right| < C_1, \forall \ \psi \in S \text{ with } \|\psi\|_{L^2} \leq 1$$

or, equivalently,

$$|\langle(1 + |\xi|^2)^{-s/2} \hat{f}, \hat{\psi}\rangle| \leq C_1, \forall \psi \in S \text{ with } \|\psi\|_{L^2} \leq 1.$$

Hence, it follows that $(1 + |\xi|^2)^{-s/2} \hat{f} \in L^2$, which proves that $f \in H^{-s}$.

Conversely, suppose that $f \in H^{-s}$ and define

$$\langle f, \phi\rangle = \int_{\underline{R}^n} (1 + |\xi|^2)^{-s/2} \hat{f}(\xi)(1 + |\xi|^2)^{s/2} \hat{\phi}(\xi) \, d\xi$$

for all $\phi \in S$. From Holder's inequality it follows that

$$|\langle f, \phi\rangle| \leq \|f\|_{-s}\|\phi\|_s,$$

hence $f \in (H^s)'$.

Finally, we leave to the reader the proof that the (-s)-norm is equivalent to the dual norm

$$\|f\| = \sup_{\|\phi\|_s \leq 1} |\langle f, \phi\rangle|.$$

This implies that on H^{-s} the strong topology of $(H^s)'$ coincides with the topology defined by the s-norm. Q.E.D.

The following theorem (a particular case of Theorem 5.2 when $\Omega = \underline{R}^n$ and $p = 2$) gives the structure of H^{-m} when m is a nonnegative integer. Its proof is based upon properties of the Fourier transform in L^2.

Theorem 5.7 *Let* $m \geq 0$ *be an integer. Then every* $f \in H^{-m}$ *can be represented as a finite sum of derivatives of order* $\leq m$ *of square integrable functions.*

Proof. If $f \in H^{-m}(\underline{R}^n)$ then, by definition,

$$(1 + |\xi|^2)^{-m/2} \hat{f}(\xi) \in L^2(\underline{R}^n).$$

Or, equivalently,

$$\hat{g}(\xi) = \frac{\hat{f}(\xi)}{1 + |\xi_1|^m + \cdots + |\xi_n|^m} \in L^2(\underline{R}^n).$$

Hence,

$$\hat{f}(\xi) = \hat{g}(\xi) + \sum_{j=1}^{n} |\xi_j|^m \hat{g}(\xi)$$

$$= \hat{g}(\xi) + \sum_{j=1}^{n} \xi_j^m \left(\frac{|\xi_j|^m}{\xi_j^m} \hat{g}(\xi) \right)$$

$$= \hat{g}(\xi) + \sum_{j=1}^{n} \xi_j^m \hat{g}_j(\xi)$$

where $\hat{g}_j(\xi) = (|\xi_j|^m/\xi_j^m)\hat{g}(\xi) \in L^2(\underline{R}^n)$. By Theorem 4.5 we get

$$f = g + \sum_{j=1}^{n} D_j^m g_j. \quad \text{Q.E.D.}$$

We shall now prove the so-called *Sobolev imbedding theorem.* Its proof is very simple, since we are only considering the case $\Omega = \underline{R}^n$ and $p = 2$, where we can use the Fourier transform and its properties. For more general results concerning the case $H^{m,p}(\Omega)$ the reader should refer to Lions [20].

Theorem 5.8. *If* $s > n/2$, then $H^s(\underline{R}^n) \subset C^0(\underline{R}^n)$ *with continuous imbedding.*

Proof. If $s > n/2$, it is easy to see that $(1 + |\xi|^2)^{-s/2} \in L^2(\underline{R}^n)$. Hence, for every $f \in H^s$ we have

$$\hat{f}(\xi) = (1 + |\xi|^2)^{-s/2} (1 + |\xi|^2)^{s/2} \hat{f}(\xi) \in L^1(\underline{R}^n). \qquad (5.6)$$

Therefore, by Proposition 4.2, it follows that $f \in C^0(\underline{R}^n)$. Next, suppose that $f_j \to 0$ in H^s. Then, by definition, $(1 + |\xi|^2)^{s/2} \hat{f}_j(\xi) \to 0$ in L^2 and by (5.6), $\hat{f}_j \to 0$ in $L^1(\underline{R}^n)$; hence $f_j \to 0$ in $C^0(\underline{R}^n)$ (Proposition 4.2). Q.E.D.

Corollary. If $s > (n/2) + k$, *where* k *is a nonnegative integer, then* $H^s(\underline{R}^n) \subset C^k(\underline{R}^n)$ *with continuous imbedding.*

Proof. Let $|\alpha| \le k$. If $f \in H^s$, then it is easily seen that $D^\alpha f \in H^{s-|\alpha|}$. Since $s - |\alpha| \ge s - k \ge n/2$ it follows, by the last theorem, that $D^\alpha f \in C^0, \forall |\alpha| \le k$. Q.E.D.

We now introduce the following vector spaces:

$$H^\infty(\underline{R}^n) = \bigcap_s H^s(\underline{R}^n) \text{ and } H^{-\infty}(\underline{R}^n) = \bigcup_s H^{-s}(\underline{R}^n).$$

As a consequence of the corollary of Theorem 5.8, $H^\infty(\underline{R}^n)$ is a subspace of $C^\infty(\underline{R}^n)$. Indeed, H^∞ coincides with the space $\mathcal{D}_{L^2}$ of all functions $\phi \in C^\infty(\underline{R}^n)$ such that $D^\alpha\phi \in L^2(\underline{R}^n)$ for all α, which will be described in Chapter 6. Such space is also discussed in Schwartz's book [29, Chapter VI, p. 55].

On H^∞ we define the coarsest locally convex topology for which the identity maps $H^\infty \to H^s, \forall s$, are continuous. On $H^{-\infty}$ we define the inductive limit topology of H^s, $\forall s$. We then have

$$H^{\infty} \subset \cdots H^s \subset H^t \subset \cdots \subset H^{-\infty} \quad (s \geq t)$$

with continuous imbeddings.

As an easy consequence of Theorem 5.7 we get the following result on the structure of $H^{-\infty}$.

Theorem 5.9. *A distribution* f *belongs to* $H^{-\infty}(\underline{R}^n)$ *if and only if it is a finite sum of derivatives of functions of* $L^2(\underline{R}^n)$.

Examples. 1. The Dirac measure δ belongs to H^s for all $s < -n/2$. In fact, we have $\hat{\delta} = 1$ and $(1 + |\xi|^2)^{s/2} \in L^2(\underline{R}^n)$ whenever $s < -n/2$.

2. If m is a positive integer such that $-m < -n/2$, it follows by Theorem 5.7 that the Dirac measure δ can be represented as a finite sum of derivatives of order $\leq m$ of functions belonging to $L^2(\underline{R}^n)$.

3. Let

$$\alpha_\varepsilon(x) = \varepsilon^{-n}\alpha\left(\frac{x}{\varepsilon}\right)$$

be the test function defined in Chapter 1, Section 1. As we know, $\alpha_\varepsilon \to \delta$ in $E'(\underline{R}^n)$ as $\varepsilon \to 0$ (remark following corollary of Theorem 3.3). Hence, $\hat{\alpha}_\varepsilon \to 1$ in S' as $\varepsilon \to 0$. But we can prove a stronger result. In fact, write

$$\hat{\alpha}_\varepsilon(\xi) = \varepsilon^{-n}\int_{\underline{R}^n} e^{-i\langle x,\xi\rangle}\alpha\left(\frac{x}{\varepsilon}\right)dx = \int_{\underline{R}^n} e^{-i\langle y,\varepsilon\xi\rangle}\alpha(y)\,dy = \hat{\alpha}(\varepsilon\xi).$$

Letting $\varepsilon \to 0$, we get that

$$\hat{\alpha}_\varepsilon(\xi) \to 1 \text{ pointwise in } \underline{R}^n.$$

If $s < -n/2$ we can write

$$\|\alpha_\varepsilon - \delta\|_s^2 = \int_{\underline{R}^n} (1 + |\xi|^2)^s |\hat{\alpha}(\varepsilon\xi) - 1|^2 \, d\xi.$$

Since the functions appearing inside the integral converge pointwise to zero as $\varepsilon \to 0$ and they are all bounded by an integrable function, then

$$\|\alpha_\varepsilon - \delta\|_s^2 \to 0 \text{ as } \varepsilon \to 0,$$

by the Lebesgue dominated convergence theorem. As a conclusion, for *all* $s < -n/2$, (α_ε) *converges to* δ *in* H^s.

4. *A fundamental solution of* $1 - \Delta$. Let Δ be the *Laplace operator*

$$\sum_{j=1}^{n} \frac{\partial^2}{\partial x_j^2}.$$

We want to find a tempered distribution E such that $(1-\Delta)E = \delta$. By taking Fourier transforms we get $(1 + |\xi|^2)\cdot\hat{E} = 1$; hence

$$\hat{E} = \frac{1}{1 + |\xi|^2}$$

is such that

$$(1 + |\xi|^2)^{(s+2)/2} \cdot \hat{E} \in L^2(\underline{R}^n), \forall\, s < -\frac{n}{2}.$$

Therefore, *the operator* $1-\Delta$ *possesses a fundamental solution* $E \in H^{s+2}(\underline{R}^n)$ *with* $s < -n/2$.

Similarly, there is a fundamental solution E of the operator $(1-\Delta)^k$ belonging to H^{s+2k}. By Theorem 5.8, we can choose k to be an integer sufficiently large that $(1-\Delta)^k$ possesses a fundamental solution belonging to $C^0(\underline{R}^n)$.

3. MULTIPLICATION AND CONVOLUTION OPERATIONS IN $H^s(\underline{R}^n)$.

Theorem 5.10 *If* $\phi \in S$ *and* $f \in H^s$ *the product* ϕf *belongs to* H^s. *Furthermore, the bilinear map*

$$S \times H^s \ni (\phi, f) \to \phi f \in H^s$$

is separately continuous.

The proof of this theorem is based upon the following two lemmas.

Lemma 5.1. *Let* $K(x, y)$ *be a continuous function on* $\underline{R}^n \times \underline{R}^n$ *and suppose that there is a constant* $C > 0$ *such that*

$$\int_{\underline{R}^n} |K(x, y)|\, dx \le C \text{ uniformly on } y$$

and

$$\int_{\underline{R}^n} |K(x, y)|\, dy \le C \text{ uniformly on } x.$$

Then

$$Af(x) = \int_{\underline{R}^n} K(x, y)f(y)\, dy$$

defines a continuous linear operator from $L^2(\underline{R}^n)$ *into* $L^2(\underline{R}^n)$.

Proof. For all $f, g \in L^2(\underline{R}^n)$ we have

$$|(Af, g)| = \left|\int_{\underline{R}^n} Af(x)\overline{g(x)}\, dx\right| = \left|\int_{\underline{R}^n \times \underline{R}^n} K(x, y)f(y)\overline{g(x)}\, dx\, dy\right|$$

$$\le \int_{\underline{R}^n \times \underline{R}^n} |K(x, y)|^{1/2}\, |f(y)|\, |K(x, y)|^{1/2}\, |g(x)|\, dx\, dy$$

$$\leq \left(\iint_{\underline{R}^n \times \underline{R}^n} |K(x, y)| \; |f(y)|^2 dx \; dy \right)^{1/2}$$

$$\cdot \left(\iint_{\underline{R}^n \times \underline{R}^n} |K(x, y)| \; |g(x)|^2 \; dx \; dy \right)^{1/2}$$

$$\leq C \cdot \|f\|_{L^2} \cdot \|g\|_{L^2} \tag{5.7}$$

which proves that $A: L^2 \to L^2$ is a continuous linear operator. Q.E.D.

Remark. It follows immediately from inequality (5.7) that the norm of the operator A is at most equal to C.

Lemma 5.2. *(Peetre's inequality). For every real number* t *we have*

$$\left(\frac{1 + |\xi|^2}{1 + |\eta|^2} \right)^t \leq 2^{|t|} \; (1 + |\xi - \eta|^2)^{|t|} \tag{5.8}$$

where $\xi, \eta \in \underline{R}^n$.

Proof. For all $\xi, \eta \in \underline{R}^n$, we have

$$1 + |\xi - \zeta|^2 = 1 + |\xi|^2 - 2\xi \cdot \zeta + |\zeta|^2$$

$$\leq 1 + 2|\xi|^2 + 2|\zeta|^2 \leq 2(1 + |\xi|^2)(1 + |\zeta|^2).$$

By setting $\eta = \xi - \zeta$ we get

$$1 + |\eta|^2 \leq 2(1 + |\xi|^2)(1 + |\xi - \eta|^2).$$

If $t < 0$, it suffices to raise both sides of the last inequality to the power $-t$ to get (5.8). If $t > 0$, by interchanging ξ and η and raising the inequality to the power t we again get (5.8). Q.E.D.

Proof of Theorem 5.10. 1. Let $\phi \in S$ and $f \in H^s$. By Problem 19 of Chapter 4, we have

$$\widehat{\phi f}(\xi) = (2\pi)^{-n}(\hat{\phi} * \hat{f})(\xi) = (2\pi)^{-n} \int_{\underline{R}^n} \hat{\phi}(\xi - \eta)\hat{f}(\eta)\, d\eta.$$

In order to prove that $\phi f \in H^s$ it suffices, by Definition 5.3, to prove that

$$(1 + |\xi|^2)^{s/2}\, \widehat{\phi f}(\xi) \in L^2(\underline{R}^n).$$

Let us write

$$(1 + |\xi|^2)^{s/2}\, \widehat{\phi f}(\xi) = (2\pi)^{-n} \int_{\underline{R}^n} (1 + |\xi|^2)^{s/2}\hat{\phi}(\xi - \eta)\hat{f}(\eta)\, d\eta$$

$$= (2\pi)^{-n} \int_{\underline{R}^n} \left(\frac{1 + |\xi|^2}{1 + |\eta|^2}\right)^{s/2} \hat{\phi}(\xi - \eta)(1 + |\eta|^2)^{s/2}\, \hat{f}(\eta)\, d\eta$$

and set

$$K(\xi, \eta) = \left(\frac{1 + |\xi|^2}{1 + |\eta|^2}\right)^{s/2} \hat{\phi}(\xi - \eta).$$

By Peetre's inequality, we get

$$|K(\xi, \eta)| \leq 2^{|s|/2}\, (1 + |\xi - \eta|^2)^{|s|/2}\, |\hat{\phi}(\xi - \eta)|$$

$$= 2^{|s|/2}(1 + |\xi - \eta|^2)^{(|s|/2)+n}\, |\hat{\phi}(\xi - \eta)|(1 + |\xi - \eta|^2)^{-n}$$

$$\leq C(1 + |\xi - \eta|^2)^{-n} \tag{5.9}$$

because $\phi \in S$. Hence, $K(\xi, \eta)$ satisfies the assumptions of Lemma 5.1 and consequently $(1 + |\xi|^2)^{s/2}\, \widehat{\phi f}(\xi) \in L^2(\underline{R}^n)$.

2. It is easy to get the following inequality:

$$\|\phi f\|_s \leq C\|f\|_s$$

which implies the continuity of the map $(\phi, f) \to \phi f$ with respect to $f \in H^s$. Next, suppose that $\phi_j \to 0$ in S and let

$$C_j = \sup_{\zeta \in \underline{R}^n} (1 + |\zeta|^2)^{(|s|/2)+n} |\hat{\phi}_j(\zeta)|$$

be the corresponding constant appearing in (5.9). By the remark following the proof of Lemma 5.1 we get

$$\|\phi_j f\|_s = \|(1 + |\xi|^2)^{s/2} \widehat{\phi_j f}(\xi)\|_{L^2} \leq C_j C' \|f\|_s.$$

Since $\phi_j \to 0$, hence $C_j \to 0$, the last inequality implies the continuity of the product ϕf with respect to $\phi \in S$. Q.E.D.

A very useful consequence of Theorem 5.10 is the following.

Corollary. Let $P = P(x, D) = \sum_{|\alpha| \leq m} a_\alpha(x) D^\alpha$ *be a partial differential operator of order* $\leq m$ *and coefficients belonging to* $S(\underline{R}^n)$. *For every real number* s, P *defines a continuous linear map from* $H^s(\underline{R}^n)$ *into* $H^{s-m}(\underline{R}^n)$.

Proof. It suffices to observe that D^α maps H^s continuously into $H^{s-|\alpha|}$ and to apply Theorem 5.10. Q.E.D.

Theorem 5.11. *If* $\phi \in S$ *and* $f \in H^s$, *the convolution* $\phi * f \in H^s$ *and the bilinear map*

$$(S \times H^s) \ni (\phi, f) \to \phi * f \in H^s$$

is separately continuous. Furthermore, $\phi * f \in H^\infty$.

Proof. By Theorem 4.9, if $\phi \in S$ and $f \in H^s$, then $\phi * f$ is a

C^∞ function. By assumption, $(1 + |\xi|^2)^{s/2} \hat{f}(\xi) \in L^2$ and since $\hat{\phi} \in S$ it follows that $(1 + |\xi|^2)^{s/2}\hat{f}(\xi)\hat{\phi}(\xi) \in L^2$; hence $\phi*f \in H^s$.

The inequality

$$\|\phi*f\|_s = \left\{\int_{\underline{R}^n} (1 + |\xi|^2)^s |\hat{f}(\xi)|^2 |\hat{\phi}(\xi)|^2 \, d\xi\right\}^{1/2}$$

$$\leq \sup_{\xi\in\underline{R}^n} |\hat{\phi}(\xi)| \cdot \|f\|_s$$

implies the separate continuity of the convolution product $\phi*f$.

Finally, we know that if $\phi \in S$, for every real number k, $(1 + |\xi|^2)^{k/2} \phi \in S$. Hence,

$$(1 + |\xi|^2)^{(s+k)/2}\hat{\phi}(\xi)\hat{f}(\xi) \in L^2$$

for every k, which implies that $\phi*f \in H^\infty$. Q.E.D.

Definition 5.4. *We say that a continuous linear operator* $L: C_c^\infty(\underline{R}^n) \to C^\infty(\underline{R}^n)$ *is of order* $\leq m$ *if it can be extended to a continuous linear operator from* $H^s(\underline{R}^n)$ *into* $H^{s-m}(\underline{R}^n)$. *We say that* L *has order equal to* m *if* m *is the* g.ℓ.b. *of all the numbers* r *such that* L *has order* $\leq r$.

Examples. 1. A partial differential operator of order $\leq m$ and coefficients belonging to $S(\underline{R}^n)$ defines, by the corollary of Theorem 5.10, an operator of order $\leq m$, according to Definition 5.4. The order is m if the partial differential operator has order m.

2. Let $\phi \in S$ and define $M_\phi: C_c^\infty \to C^\infty$ by

$$M_\phi(f) = \phi \cdot f.$$

By Theorem 5.10, the *multiplication operator* M_ϕ has order ≤ 0.

3. Let $\phi \in S$ and define $L_\phi: C_c^\infty \to C^\infty$ by

$$L_\phi(f) = \phi*f.$$

By Theorem 5.11, the *convolution operator* L_ϕ has order $-\infty$.

PROBLEMS

1. Prove that, if $m \geq \ell$, $H^{m,p}(\Omega) \subset H^{\ell,p}(\Omega)$ with continuous imbedding.

2. Show that the coarsest topology on $H^{m,p}(\Omega)$ for which the maps

$$D^\alpha: H^{m,p}(\Omega) \to L^p(\Omega), \quad |\alpha| \leq m,$$

are continuous coincides with the one defined by the norm (5.1).

3. Let Ω be an open subset of $\underline{R}^n$ and let $f \in H_0^{m,p}(\Omega)$. Prove that the function

$$\tilde{f} = \begin{cases} f \text{ on } \Omega \text{ a.e.} \\ 0 \text{ on } \underline{R}^n - \Omega \end{cases}$$

belongs to $H^{m,p}(\underline{R}^n)$.

4. Let Ω be the unit ball and let f be the function identically equal to one on Ω. Prove that: (i) $f \in H^{1,p}(\Omega)$; (ii) the function $\tilde{f}$ does not belong to $H^{1,p}(\underline{R}^n)$. Conclude that $f \in H^{1,p}(\Omega)$ but $f \notin H_0^{1,p}(\Omega)$.

5. Show that the linear map J defined in the proof of Theorem 5.2 is an isometry from $H^{m,p}(\Omega)$ into $(L^p(\Omega))^N$.

6. Let $f \in S'(\underline{R}^n)$. Prove that the following are equivalent conditions:

(i) $D^\alpha f \in L^2(\underline{R}^n)$, $\forall\, |\alpha| \leq m$;

(ii) $\xi^\alpha \hat{f}(\xi) \in L^2(\underline{R}^n)$, $\forall\, |\alpha| \leq m$;

(iii) $P(\xi)\hat{f}(\xi) \in L^2(\underline{R}^n)$ for all polynomials $P(\xi)$ with constant coefficients and degree $\leq m$;

(iv) $(1 + |\xi|^2)^{m/2}\hat{f}(\xi) \in L^2(\underline{R}^n)$;

(v) $(1 + |\xi|)^m\hat{f}(\xi) \in L^2(\underline{R}^n)$.

7. Show that when $s = m$, a nonnegative integer, then Definition 5.3 coincides with Definition 5.1 and that the norms (5.1) and (5.5) are equivalent.

8. Let s be a real number. Prove that the imbeddings $\mathcal{S} \to H^s$ and $H^s \to \mathcal{S}'$ are continuous.

9. Complete the proof Theorem 5.6 by showing that the (-s)-norm is equivalent to the dual norm.

10. Show that D^α defines a continuous linear map from H^s into $H^{s-|\alpha|}$.

Chapter 6

ON SOME SPACES OF DISTRIBUTIONS

1. THE SPACES $\mathcal{D}_{L^p}$ AND THEIR DUALS

Following Schwartz [28, Chapter VI, p. 55], we make the following definition.

Definition 6.1. *Let* p *be a real number such that* $1 \leq p \leq +\infty$. *We denote by* $\mathcal{D}_{L^p}(\underline{R}^n)$ *the space of all functions* $\phi \in C^\infty(\underline{R}^n)$ *such that* $\partial^\alpha\phi \in L^p(\underline{R}^n)$ *for all* $\alpha \in \underline{N}^n$ *equipped with the coarsest locally convex topology for which the maps*

$$\partial^\alpha : \mathcal{D}_{L^p}(\underline{R}^n) \to L^p(\underline{R}^n)$$

are continuous for all $\alpha \in \underline{N}^n$.

The topology of $\mathcal{D}_{L^p}$ coincides trivially with the one defined by the countable family of norms

$$\|\phi\|_{m,p} = \left(\sum_{|\alpha| \leq m} \|\partial^\alpha\phi\|_{L^p}^p \right)^{1/p}, \quad m \in \underline{N}. \tag{6.1}$$

It is obvious that $\mathcal{D}_{L^p}$ is a *Frechet space*.

We have the following imbeddings:

$$C_c^\infty(\underline{R}^n) \subset \mathcal{D}_{L^p}(\underline{R}^n) \subset \mathcal{D}'(\underline{R}^n), \; 1 \leq p < +\infty,$$

with continuous injections. Furthermore, if $1 \leq p < +\infty$, C_c^∞ is a dense subspace of $\mathcal{D}_{L^p}$; hence $\mathcal{D}_{L^p}$ is a *normal space of distribution* in $\underline{R}^n$. However, C_c^∞ is *not* dense in $\mathcal{D}_{L^\infty}$. (Proofs are left to the reader.) We then define $\dot{\mathcal{D}}_{L^\infty}$ as the subspace of all functions in $\mathcal{D}_{L^\infty}$ converging to zero at infinity. It is clear that $\dot{\mathcal{D}}_{L^\infty}$ is a *closed* subspace of $\mathcal{D}_{L^\infty}$; hence it is a Frechet space. We also have $C_c^\infty \subset \mathcal{D}_{L^\infty} \subset \mathcal{D}'$ with continuous injections and C_c^∞ dense in $\dot{\mathcal{D}}_{L^\infty}$; therefore, $\dot{\mathcal{D}}_{L^\infty}$ is a normal space of distributions.

In order to have further properties on the structure of the spaces $\mathcal{D}_{L^p}$, we need the following theorem.

Theorem 6.1. *Let* A *and* a *be real numbers such that* $0 < a < A$ *and denote by* B_A *and* B_{A-a} *concentric balls of radius* A *and* A - a, *respectively. If a distribution* $T \in \mathcal{D}'(\underline{R}^n)$ *is such that all its derivatives of rank*[1] ≤ 1 *are functions such that their* L^p *norms on* B_A *are bounded by a constant* M, *then on* B_{A-a}, T *is function bounded by* $C(A, n)a^{-n}M$, *where* C(A, n) *is a constant independent of* T *and of* a.

Proof. By corollary 3 of Theorem 1.1, there is a function $\psi \in C_c^\infty(\underline{R}^n)$, such that $\operatorname{supp} \psi \subset B_A$, $\psi(x) = 1$, on B_{A-a} and for all $\alpha \in \underline{N}^n$,

$$|\partial^\alpha \psi(x)| \leq C(\alpha, n)\cdot a^{-|\alpha|}.$$

On the other hand, let

$$Y(x) = \begin{cases} 1 \text{ if } x_1 \geq 0, \cdots, x_n \geq 0 \\ \\ 0 \text{ otherwise} \end{cases}$$

be the Heaviside function in $\underline{R}^n$. It is a fundamental solution of $\partial^n/\partial x_1 \cdots \partial x_n$ (Chapter 2, Section 5, Example 6). We can write

[1] We recall that if $\alpha \in \underline{N}^n$, rank $\alpha = \max_{1 \leq j \leq n} (\alpha_j)$.

$$\psi T = \delta * (\psi T) = \frac{\partial^n Y}{\partial x_1 \cdots \partial x_n} * (\psi T) = Y * \frac{\partial^n (\psi T)}{\partial x_1 \cdots \partial x_n};$$

hence T is a function in B_{A-a}. For all $x = (x_1, \cdots, x_n) \in B_{A-a}$, write

$$T(x) = (\psi T)(x) = \int_{-\infty}^{x_1} \cdots \int_{-\infty}^{x_n} \frac{\partial^n (\psi T)}{\partial y_1 \cdots \partial y_n} \, dy.$$

By Leibniz's formula,

$$\frac{\partial^n (\psi T)}{\partial y_1 \cdots \partial y_n} = \sum_{\substack{|\alpha|+|\beta|=n \\ \mathrm{rank}\,\alpha \le 1 \\ \mathrm{rank}\,\beta \le 1}} C_{\alpha,\beta} \frac{\partial^\alpha \psi}{\partial y^\alpha} \cdot \frac{\partial^\beta T}{\partial y^\beta}.$$

By Hölder's inequality, we get

$$\int_{-\infty}^{x_1} \cdots \int_{-\infty}^{x_n} \left|\frac{\partial^\alpha \psi}{\partial y^\alpha}\right| \cdot \left|\frac{\partial^\beta T}{\partial y^\beta}\right| dy \le \int \underset{B_A}{\cdots} \int \left|\frac{\partial^\alpha \psi}{\partial y^\alpha}\right| \cdot \left|\frac{\partial^\beta T}{\partial y^\beta}\right| dy$$

$$\le \left(\int \underset{B_A}{\cdots} \int \left|\frac{\partial^\alpha \psi}{\partial y^\alpha}\right|^{p'} dy\right)^{1/p'}$$

$$\cdot \left(\int \underset{B_A}{\cdots} \int \left|\frac{\partial^\beta T}{\partial y^\beta}\right|^{p} dy\right)^{1/p}$$

$$\le C(\alpha, n) a^{-|\alpha|} M$$

which implies the desired inequality. Q.E.D.

Corollary 1. *If* $T \in \mathcal{D}'(\underline{R}^n)$ *is such that all its derivatives of rank* ≤ 1 *are functions belonging to* $L^p(\underline{R}^n)$, *then we have the following estimate*

$$\sup_{x \in \underline{R}^n} |T(x)| \le C \cdot \sum_{\mathrm{rank}\,\alpha \le 1} \|\partial^\alpha T\|_{L^p}.$$

Proof. It suffices to apply the theorem to balls with center x, a variable element in $\underline{R}^n$, and radius 2 and 1. Q.E.D.

Corollary 2. If $\phi \in \mathcal{D}_{L^p}$ $(1 \le p \le +\infty)$ *then* ϕ *is bounded in* $\underline{R}^n$ *Moreover, if* $1 \le p < +\infty$, ϕ *converges to zero at infinity.*

Proof. The boundedness of $\phi \in \mathcal{D}_{L^p}$ follows from Corollary 1. If $\phi \in \mathcal{D}_{L^p}$, $1 \le p < +\infty$, it is possible to find a number $R > 0$ so large that

$$\int \cdots \int_{|x|>R} \left| \frac{\partial^\alpha \phi}{\partial x^\alpha} \right| dx < \varepsilon$$

for all α with rank ≤ 1. By applying Theorem 6.1, we get that $|\phi(x)|$ is arbitrarily small for large $|x|$. Q.E.D.

Corollary 3. If $1 \le p \le q \le +\infty$, *then* $\mathcal{D}_{L^p} \subset \mathcal{D}_{L^q}$ *with continuous imbedding.*

Proof. When $q = +\infty$, the result is a consequence of Corollary 2. Suppose $q < +\infty$. By setting $\phi^{(\beta)} = \partial^\beta \phi$ and applying Corollary 1, we get

$$\int_{\underline{R}^n} |\alpha^{(\beta)}(x)|^q \, dx \le \sup_{x \in \underline{R}^n} |\phi^{(\beta)}(x)|^{q-p} \int_{\underline{R}^n} |\phi^{(\beta)}(x)|^p \, dx$$

$$\le C \sum_{\text{rank}\,\alpha \le 1} \| \partial^\alpha \phi^{(\beta)} \|_{L^p}^{q-p} \cdot \| \phi^\beta \|_{L^p}^p . \quad \text{Q.E.D.}$$

Remark. The spaces $\mathcal{D}_{L^p}$, $1 \le p \le \infty$, are related to the Soblev spaces $H^{m,p}(\underline{R}^n)$. Indeed, by the Sobolev imbedding theorem, $H^{m,p}(\underline{R}^n)$ is a space of continuous functions provided that m is sufficiently large. Hence, it follows that

$$\mathcal{D}_{L^p}(\underline{R}^n) = \bigcap_{m=0}^{\infty} H^{m,p}(\underline{R}^n)$$

and the above-defined topology of $\mathcal{D}_{L^p}$ coincides with the coarsest one for which the imbeddings $\mathcal{D}_{L^p} \to H^{m,p}$ are continuous.

2. THE DUAL OF $\mathcal{D}_{L^p}$

As we have remarked above, the spaces $\mathcal{D}_{L^p}$, $1 \leq p + \infty$, and $\mathcal{D}_{L^\infty}$ are normal spaces of distributions; therefore their duals are subspaces of $\mathcal{D}'(\underline{R}^n)$.

Definition 6.2. *We denote by* $\mathcal{D}'_{L^p}$, $1 < p \leq +\infty$, *the dual of* $\mathcal{D}_{L^{p'}}$, *where* $(1/p) + (1/p') = 1$. *We denote by* $\mathcal{D}'_{L^1}$ *the dual of* $\mathcal{D}_{L^\infty}$.

These duals are subspaces of $\mathcal{D}'(\underline{R}^n)$. As in previous cases discussed before, we can define the strong topology of $\mathcal{D}'_{L^p}$ and prove that the imbedding $\mathcal{D}'_{L^p} \to \mathcal{D}'$ is strongly continuous.

In view of the reflexivity of L^p, $1 < p < +\infty$, it follows that the spaces $\mathcal{D}_{L^p}$ are reflexive. Also, it can be shown that the daul of $\mathcal{D}'_{L^1}$ is $\mathcal{D}_{L^\infty}$[28, Chapter VI, Section 8].

Since

$$\mathcal{D}_{L^{q'}} \subset \mathcal{D}_{L^{p'}}, \forall\, q' \leq p',$$

$C_c^\infty(\underline{R}^n)$ is dense in $\mathcal{D}_{L^{q'}}$, $1 \leq q' < +\infty$, and in $\mathcal{D}_{L^\infty}$, it follows that

$$\mathcal{D}'_{L^p} \subset \mathcal{D}'_{L^q}, \ \forall\, p \leq q.$$

The next theorem, closely related to Theorem 5.2, gives us the structure of $\mathcal{D}'_{L^p}$.

Theorem 6.2. *A distribution* $T \in \mathcal{D}'_{L^p}$ *if and only if there is an integer* $m = m(T) > 0$ *such that*

$$T = \sum_{|\alpha| \leq m} \partial^\alpha f_\alpha \tag{6.2}$$

with $f_\alpha \in L^p$.

Proof. As we shall see the proof is very similar to that of Theorem 5.2. Consider first the case $1 < p \leq +\infty$. If $T \in \mathcal{D}'(\underline{R}^n)$ is of the form (6.2) then T defines a continuous linear functional on $\mathcal{D}_{L^{p'}}$; hence $T \in \mathcal{D}'_{L^p}$.

Conversely, suppose that $T \in \mathcal{D}'_{L^p}$. Then there is a neighborhood of zero $W(m, \delta)$ in $\mathcal{D}_{L^{p'}}$ such that

$$|\langle T,\phi\rangle| \leq 1, \forall\, \phi \in W(m,\delta).$$

It then follows that there is a constant $C > 0$ such that

$$|\langle T,\phi\rangle| \leq C \sup_{|\alpha|\leq m} \| \partial^\alpha \phi \|_{L^{p'}}, \forall\, \phi \in \mathcal{D}_{L^{p'}}. \tag{6.3}$$

If N denotes the number of n-tuples α such that $|\alpha| \leq m$, set $(L^{p'})^N = L^{p'} \times\cdots\times L^{p'}$ (N times) and define the map

$$J: \mathcal{D}_{L^p} \ni \phi \to (\partial^\alpha \phi)_{|\alpha|\leq m} \in (L^{p'})^N.$$

Since J is obviously a one-to-one map, we can identify $J\mathcal{D}_{L^{p'}}$ with a proper subspace of $(L^{p'})^N$. On $J\mathcal{D}_{L^{p'}}$ we define the linear functional $F: J\mathcal{D}_{L^{p'}} \to \underline{C}$ as follows

$$F\left((\partial^\alpha \phi)_{|\alpha|\leq m}\right) = \langle T,\phi\rangle, \forall\, \phi \in \mathcal{D}_{L^{p'}}.$$

By (6.3), F is a continuous linear functional on $J\mathcal{D}_{L^{p'}}$ equipped with the topology induced by $(L^{p'})^N$. By the Hahn-Banach theorem, F extends to a continuous linear functional on $(L^{p'})^N$. Since the dual of the product $(L^{p'})^N$ can be identified with the product $(L^p)^N$, there are functions $g_\alpha \in L^p$, $|\alpha| \leq m$, such that

$$\langle T,\phi\rangle = F\left((\partial^\alpha \phi)_{|\alpha|\leq m}\right) = \sum_{|\alpha|\leq m} \int_{\underline{R}^n} g_\alpha \frac{\partial^\alpha \phi}{\partial x^\alpha}\, dx;$$

for all $\phi \in \mathcal{D}_{L^{p'}}$; hence

$$T = \sum_{|\alpha| \le m} \partial^\alpha f_\alpha$$

with $f_\alpha = (-1)^{|\alpha|} g_\alpha$. Q.E.D.

The case $p = 1$ requires only a slight modification of the previous proof. We leave the details to the reader.

As we mentioned in our remark following Corollary 3 of Theorem 6.1, the space $\mathcal{D}_{L^2}$ coincides, in the algebraic and topological senses, with H^∞. Its dual $\mathcal{D}'_{L^2}$ will then coincide with $H^{-\infty}$. Thus, Theorem 5.9 is a consequence of Theorem 6.2

By using a fundamental solution of the operator $(1 - \Delta)^k$ with sufficiently large k, we can improve the result of Theorem 6.2 in the following way.

Corollary. For every distribution $T \in \mathcal{D}'_{L^p}$ *there is an integer* $m = m(T) > 0$ *such that*

$$T = \sum_{|\alpha| \le m} \partial^\alpha f_\alpha$$

where the f_α *are bounded continuous functions belonging to* L^p. *Moreover, if* $p < +\infty$, *each* f_α *converges to zero at infinity.*

Proof. By Theorem 6.2, $T = \Sigma_{|\beta| \le \ell} \partial^\beta g_\beta$ with $g_\beta \in L^p$. Let E be a fundamental solution of $(1 - \Delta)^k$ and choose k so large that E is a continuous function (see Chapter 5, Section 2, Example 4). Let $\gamma \in C_c^\infty(\underline{R}^n)$ be such that γ is equal to one on a neighborhood of the origin in $\underline{R}^n$ and set $F = \gamma E$. We have

$$(1 - \Delta)^k F = \delta - \phi, \tag{6.4}$$

where $\phi \in C_c^\infty(\underline{R}^n)$. The distribution F is called a *parametrix* of the partial differential operator $(1 - \Delta)^k$. From (6.4) it follows that every distribution $g \in \mathcal{D}'(\underline{R}^n)$ can be written as follows:

$$g = (1 - \Delta)^k(\gamma E * g) + \phi * g. \tag{6.5}$$

Moreover, if $g \in L^p$, since γE and ϕ are continuous functions with compact support, it follows (see Theorem 1.1) that $\gamma E*g$ and $\phi*g$ are bounded continuous functions belonging to L^p and converging to zero at infinity if $p < +\infty$. Hence, every $g \in L^p$ can be written as a finite sum of derivatives (in the sense of distributions) of bounded continuous functions belong to L^p and converging to zero at infinity if $p < +\infty$. Applying these facts to the functions g_β appearing in the representation of T, we obtain the desired result. Q.E.D.

3. FOURIER TRANSFORMS

We shall prove a few results concerning the Fourier transform of elements of $\mathcal{D}'_{L^p}$ in the cases p = 1 and p = 2.

Theorem 6.3. *If* $\phi \in \mathcal{D}_{L^1}$ *its Fourier transform is a continuous function rapidly decreasing at infinity. If* $T \in \mathcal{D}'_{L^1}$, *its Fourier transform is a continuous function slowly increasing at infinity.*

Proof. 1. Suppose that $\phi \in \mathcal{D}_{L^1}$. Then for all $\alpha \in \underline{N}^n$, $D^\alpha\phi \in L^1$; hence, by Proposition 4.2, $\xi^\alpha\hat{\phi}(\xi)$ is a continuous bounded function on $\underline{R}^n$. This implies that for every polynomial $P(\xi)$, the product $P(\xi)\hat{\phi}(\xi)$ is a bounded function on $\underline{R}^n$; hence $\hat{\phi}$ is a continuous function rapidly decreasing at infinity.

2. If $T \in \mathcal{D}'_{L^1}$ then $T = \Sigma_{|\alpha|\leq m} D^\alpha f_\alpha$, where the f_α are bounded continuous functions belonging to $L^1(\underline{R}^n)$. By proposition 4.2, $\hat{T}(\xi) = \Sigma_{|\alpha|\leq m}\, \xi^\alpha \hat{f}_\alpha$, where the $\hat{f}_\alpha$ are bounded continuous functions. Hence, $\hat{T}(\xi)$ can be written as a product of a polynomial and a bounded continuous function. Q.E.D.

Theorem 6.4. *A distribution* T *belongs to* $\mathcal{D}'_{L^2}$ *if and only if its Fourier transform is the product of a polynomial and an* L^2 *function.*

Proof. If $T \in \mathcal{D}'_{L^2}$, then by Theorem 6.2, $T = \Sigma_{|\alpha|\leq m}\, D^\alpha f_\alpha$, with

$\hat{f}_\alpha \in L^2(\underline{R}^n)$. Hence,

$$T(\xi) = \sum_{|\alpha|\le m} \xi^\alpha \hat{f}_\alpha,$$

with $\hat{f}_\alpha \in L^2$ by Theorem 4.5. Since the function

$$\hat{h}(\xi) = \frac{\sum_{|\alpha|\le m} \xi^\alpha \hat{f}_\alpha}{(1+|\xi|^2)^{m/2}}$$

belongs to L^2, we get $\hat{T}(\xi) = (1+|\xi|^2)^{m/2}\,\hat{h}(\xi)$, which implies the desired result. The converse is also trivial. Q.E.D.

Remark. Theorem 6.4 can also be easily derived from Theorems 5.7 and 5.9.

4. THE STRUCTURE OF $S'(\underline{R}^n)$.

As we know, a derivative of a continuous function slowly increasing at infinity defines a tempered distribution. We shall now prove that, conversely, every tempered distribution is the derivative of a continuous function slowly increasing at infinity. We start by proving the following lemma.

Lemma 6.1. *If* T *is a tempered distribution, there is a number* $k > 0$ *such that* $(1 + r^2)^{-k/2}\, T \in \mathcal{D}'_{L^\infty}$.

Proof. If $T \in S'$, we can find a neighborhood of zero

$$V(k,m,\varepsilon) = \left\{\phi \in S\colon\ |(1+r^2)^{k/2}\partial^p\phi(x)| \le \varepsilon, \forall\ |p| \le m, \forall\ x \in \underline{R}^n\right\}$$

in S such that $|\langle T, \phi\rangle| \le 1$, $\forall\ \phi \in V$.

We claim that there is a neighborhood of zero W in $\mathcal{D}_{L^1}$ such that for all $\psi \in S \cap W$, $\phi = (1+r^2)^{-m/2}\,\psi \in V$. Indeed, let

$$W(m+n, \eta) = \left\{\psi \in \mathcal{D}_{L^1} : \sup_{|\alpha|\le m+n} \int\left|\frac{\partial^\alpha\phi}{\partial x^\alpha}(x)\right| dx \le \eta\right\}$$

where η is to be determined later. First, we observe that for every n-tuple $s \in \underline{N}^n$ we have

$$|\partial^s(1 + r^2)^{-k/2}| \le C_{s,k}(1 + r^2)^{-k/2}.$$

Let $p \in \underline{N}^n$ be such that $|p| \le m$. By the Leibniz formula,

$$\partial^p\phi = \sum_{s+q=p} \frac{p!}{s!q!} \partial^s[1 + r^2)^{-k/2}] \partial^q\psi;$$

hence

$$|\partial^p\phi| \le C_{p,k}(1 + r^2)^{-k/2} \sum_{q\le p} |\partial^p\psi|.$$

On the other hand,

$$\partial^q\psi(x) = \int_{-\infty}^{x_1} \cdots \int_{-\infty}^{x_n} \frac{\partial^n}{\partial t_1 \cdots \partial t_n}\left[\frac{\partial^q\psi}{\partial t^q}(t)\right] dt_1 \cdots dt_n$$

which yields

$$\sup_{x\in R^n} |\partial^q\psi(x)| \le \left\|\frac{\partial^n}{\partial t_1 \cdots \partial t_n}\left(\frac{\partial^q\psi}{\partial t^q}\right)\right\|_{L^1} \le \eta$$

since $|q| + n \le m + n$. Making the appropriate replacement, we get

$$(1 + r^2)^{k/2}|\partial^p\phi| \le C'_{p,k}\cdot\eta.$$

Choosing η so that $C'_{p,k}\cdot\eta \le \varepsilon$ we see that $\phi \in V$ whenever $\psi \in \mathcal{S} \cap W$.

Therefore, for all $\psi \in \mathcal{T} \cap W$ we have

$$|<(1 + r^2)^{-k/2} T, \psi>| = |<T, (1 + r^2)^{-k/2} \psi>| \le 1$$

which implies, since $\mathcal{S}$ is dense in $\mathcal{D}_{L^1}$, that $(1 + r^2)^{-k/2}$ T is a continuous linear functional on $\mathcal{D}_{L^1}$. Q.E.D.

Theorem 6.5. *A distribution* $T \in \mathcal{S}'$ *if and only if it can be represented as a finite sum*

$$T = \sum \partial^\alpha f_\alpha \tag{6.6}$$

where the f_α *are continuous functions slowly increasing at infinity.*

Proof. By Lemma 6.1, there is a number $k > 0$ such that $(1 + r^2)^{-k/2}\, T \in \mathcal{D}'_{L^\infty}$. By Theorem 6.2, we can write

$$(1 + r^2)^{-k/2} T = \sum \partial^\gamma h_\gamma,\ h_\gamma \in L^\infty.$$

Set

$$g_\gamma(x) = \int_0^{x_1} \cdots \int_0^{x_n} h_\gamma(t)\, dt_1 \cdots dt_n.$$

Since

$$|g_\gamma(x)| \le |x_1| \cdots |x_n| \sup_{x \in \underline{R}^n} |h_\gamma(x)|,$$

g_γ is a continuous function slowly increasing at infinity. Making the appropriate replacement, we get

$$T = \sum (1 + r^2)^{k/2}\, \partial^\beta g_\beta.$$

Finally, noticing that every term $(1 + r^2)^{k/2} \partial^\beta g_\beta$ can be written as a finite sum

$$\sum \partial^\alpha (1 + r^2)^{k_\alpha/2}\, g'_\alpha$$

where g'_α is a bounded continuous function, our result follows at once. Q.E.D.

Since the function f_α appearing in (6.6) is a continuous function slowly increasing at infinity, it can be written as follows:

$$f_\alpha = (1 + r^2)^{\ell_\alpha/2} g_\alpha$$

where g_α is a bounded continuous function in $\underline{R}^n$.

On the other hand, we can reduce the sum (6.6) to a single derivative. Indeed, suppose that

$$T = (1 + r^2)^{k/2} f + \partial_j \left[(1 + r^2)^{\ell/2} f_j\right],$$

where f and f_j are bounded continuous functions. Set

$$g = \int_0^{x_j} (1 + r^2)^{k/2} f \, dt_j$$

and

$$h = \frac{g}{(1+r^2)^{k/2+1}} .$$

It is clear that $\partial_j g = (1 + r^2)^{k/2} f$ and that h is a bounded continuous function. Making the appropriate replacement we get

$$T = \partial_j \left[(1 + r^2)^{m/2} F\right]$$

with a suitable m, where F is a bounded continuous function. Proceeding by induction, we can derive from Theorem 6.5 the following result.

Corollary. A distribution T *belongs to* S' *if and only if it can be represented as*

$$T = \partial^\alpha[(1+r^2)^{k/2} f(x)]$$

where f(x) *is a bounded continuous function on* $\underline{R}^n$.

5. THE SPACE $\mathcal{O}'_C$ OF DISTRIBUTIONS WHICH ARE RAPIDLY DECREASING AT INFINITY

In Chapter 4, Section 8 we have seen that the elements of $\mathcal{S}$ operate on $\mathcal{S}'$ by convolution (Theorem 4.9) and also that $\mathcal{E}'$ operates on $\mathcal{S}'$ by convolution (Theorem 4.10). In both cases, the operation is separately continuous and the Fourier transform operator maps the convolution product into the product of the corresponding Fourier transforms. It is then natural to ask which is the most general space of distributions operating continuously on $\mathcal{S}'$ by convolution and such that the Fourier transform maps convolution products into the product of the corresponding Fourier transforms. Following Schwartz [28, Chapter VII, p. 100], we make the following definition.

Definition 6.3. *Let* $\mathcal{O}'_C(R^n)$ *be the space of all distributions* $T \in \mathcal{D}'(\underline{R}^n)$ *such that for all real numbers* k

$$(1 + r^2)^{k/2}T \in \mathcal{D}'^{\infty}_L. \tag{6.7}$$

In other words, $T \in \mathcal{O}'_C$ if and only if for all k, $(1 + r^2)^{k/2}T$ is a finite sum of derivatives of functions belonging to L^∞ (Theorem 6.2). Or, equivalently, $T \in \mathcal{O}'_C$ if and only if for all polynomials P(x) we have $P(x)T \in \mathcal{D}'^{\infty}_L$.

Examples. 1. Every distribution with compact support defines an element of $\mathcal{O}'_C$.

2. Every element of $\mathcal{O}'_C$ defines a tempered distribution.

In a more consistent way the space $\mathcal{O}'_C$ can be defined as a dual of a normal space of distributions, namely the space $\mathcal{O}_C$ of all C^∞ functions ϕ on $\underline{R}^n$, such that there is an integer k such that $(1 + r^2)^{k/2}\, \partial^\alpha\phi$ vanishes at infinity for all $\alpha \in \underline{N}$. The space $\mathcal{O}_C$ can be equipped with a Hausdorff locally convex topology such that it becomes a normal space of distributions whose dual consists of all distributions on $\underline{R}^n$ satisfying the condition of Definition 6.3. On $\mathcal{O}'_C$ we can define its strong topology and the imbedding $\mathcal{O}'_C \to \mathcal{D}'$ is strongly continuous. We also have $\mathcal{E}' \subset \mathcal{O}'_C$ and $\mathcal{O}'_C \subset \mathcal{S}'$ with continuous imbeddings. We are not going to prove these results here and the reader should refer to Horvath [17] for a more detailed exposition.

From definition 6.3 we can immediately derive a notion of convergence to zero in $\mathcal{O}'_C$, namely distributions $T_j \in \mathcal{O}'_C$ converge to zero if and only if for all k the distributions $(1 + r^2)^{k/2} T_j$ converge to zero in $\mathcal{D}'_{L^\infty}$. We remark that this notion corresponds to that of strong convergence to zero in $\mathcal{O}'_C$.

Theorem 6.6. *A distribution* $T \in \mathcal{O}'_C$ *if and only if for every* $k > 0$ *there is an integer* $m = m(k)$ *such that*

$$T = \sum_{|\alpha| \leq m} \partial^\alpha f_\alpha \tag{6.8}$$

where the f_α *are continuous functions such that* $(1 + r^2)^{k/2} f_\alpha \in L^\infty$.

Proof. If $T \in \mathcal{O}'_C$, then by definition

$$(1 + r^2)^{k/2} T = \sum_\beta \partial^\beta g_\beta,$$

where g_β is a bounded function. Hence,

$$T = \sum_\beta (1 + r^2)^{-k/2} \partial^\beta g_\beta$$

and by the Leibniz formula we can write

$$(1 + r^2)^{-k/2} \partial^\beta g_\beta = \partial^\beta \left((1 + r^2)^{-k/2} \quad g_\beta \right) -$$

$$\sum_{0 \leq |t| < |\beta|} c_{s,t} \partial^s \left[(1 + r^2)^{-k/2} \right] \partial^t g_\beta .$$

Let us show by induction on $|t|$ that

$$\partial^s\left[(1 + r^2)^{-k/2}\right]\partial^t g_\beta$$

is a finite sum of derivatives of continuous functions whose product by $(1 + r^2)^{k/2}$ are bounded in $\underline{R}^n$. Indeed, if $t = 0$ we set

$$h = \partial^s\left\{(1 + r^2)^{-k/2}\right\} g_\beta.$$

Since

$$|h| \le C(1 + r^2)^{-k/2}|g_\beta|$$

it follows that $(1 + r^2)^{k/2}h \in L^\infty$. Suppose that the result is true when $|t| = \ell$. Let $|t| = \ell + 1$ and write

$$\partial^s\left[(1 + r^2)^{-k/2}\right]\partial^t g_\beta = \partial_j\left\{\partial^s\left[(1 + r^2)^{-k/2}\right]\partial^{t'} g_\beta\right\}$$

$$- \partial_j\partial^s\left[(1 + r^2)^{-k/2}\right]\partial^{t'} g_\beta.$$

Hence, the left-hand side is a sum of derivatives of continuous functions whose product with $(1 + r^2)^{k/2}$ belongs to L^∞.

Conversely, suppose that a distribution T can be represented by (6.8). For every $\phi \in \mathcal{D}_{L^1}$ and every $|\alpha| \le m$ we set

$$<(1 + r^2)^{k/2}\,\partial^\alpha f_\alpha, \phi> = (-1)^{|\alpha|}\int_{\underline{R}^n} f_\alpha \partial^\alpha\left[(1 + r^2)^{k/2}\phi\right] dx.$$

Since $\partial^\alpha((1 + r^2)^{k/2}\phi)$ is bounded by a sum $(1 + r^2)^{k/2}\sum_{|\ell|\le|\alpha|} \partial^\ell\phi$ with $\partial^\ell\phi \in L^1$ and $(1 + r^2)^{k/2}f_\alpha \in L^\infty$, we get

$$|<(1 + r^2)^{k/2}\,\partial^\alpha f_\alpha, \phi>| \le C\sum_{|\ell|\le|\alpha|}\int |(1 + r^2)^{k/2}f_\alpha\partial^\ell\phi|\,dx$$

$$\le C\|(1 + r^2)^{k/2}f_\alpha\|_{L^\infty}\cdot\sum_{|\ell|\le|\alpha|}\|\partial^\ell\phi\|_{L^1}$$

hence $(1 + r^2)^{k/2} T \in \mathcal{D}'_{L^\infty}$. Q.E.D.

The following is an easy consequence of Theorem 6.6.

Corollary 1. *A distribution* T *belongs to* $\mathcal{O}'_C$ *if and only if for every* $k \geq 0$ *there is an integer* $m = m(k)$ *such that*

$$T = \sum_{|\alpha| \leq m} \partial^\alpha f_\alpha,$$

where f_α *is a continuous function on* $\underline{R}^n$ *such that*

$$\lim_{r \to +\infty} (1 + r^2)^{k/2} |f_\alpha(x)| = 0.$$

Corollary 2. *If* $T \in \mathcal{O}'_C$, *then* $T \in \mathcal{D}'_{L^1}$.

Proof. If in Theorem 6.6 we take k such that $(1 + r^2)^{-k/2}$ is integrable in $\underline{R}^n$, the functions f_α appearing in (6.8) belong to $L^1(\underline{R}^n)$ and the corollary is then a consequence of Theorem 6.2, Q.E.D.

Since $\mathcal{D}'_{L^p} \subset \mathcal{D}'_{L^q}, \forall\, p \leq q$, it follows from Corollary 2 that if $T \in \mathcal{O}'_C$ then $T \in \mathcal{D}'_{L^p}, \forall\, p$.

In order to show that the space $\mathcal{O}'_C$ operates on $\mathcal{S}'$ by convolution, we shall first prove the following lemma.

Lemma 6.2 *If* $S = \partial^\alpha\big((1 + r^2)^{k/2} f(x)\big) \in \mathcal{S}'$ *with* f(x) *a bounded continuous function on* $\underline{R}^n$ *and if* $\phi \in \mathcal{S}$ *then*

$$F(x) = \langle S_y, \phi(x + y)\rangle$$

is a C^∞ *function on* $\underline{R}^n$ *such that* $(1 + |x|^2)^{-\ell/2} F(x) \in \mathcal{D}_{L^1}$ *for all* $\ell > k + n$. *Furthermore, if* $\phi_j \to 0$ *in* $\mathcal{S}$ *then* $F_j \to 0$ *in* $\mathcal{D}_{L^1}$.

Proof. We have

$$F(x) = \langle S_y, \phi(x + y)\rangle = (-1)^{|\alpha|} \int_{\underline{R}^n} (1 + |y|^2)^{k/2} f(y) \frac{\partial^\alpha \phi}{\partial y^\alpha} (x + y)\, dy;$$

hence

$$\frac{\partial^\beta F}{\partial x^\beta} = (-1)^{|\alpha|} \int (1 + |y|^2)^{k/2} f(y) \frac{\partial^{\alpha+\beta}}{\partial x^\beta \partial y^\alpha} \phi(x + y)\, dy.$$

Since f is bounded on $\underline{R}^n$, we get, by applying Lemma 5.2,

$$\left|\frac{\partial^\beta F}{\partial x^\beta}(x)\right| \le C(1 + |x|^2)^{k/2} \int_{\underline{R}^n} (1 + |x + y|^2)^{k/2} \frac{\partial^{\alpha+\beta}}{\partial x^\beta \partial y^\alpha} \phi(x + y)\, dy$$

$$= C(1 + |x|^2)^{k/2} \int_{\underline{R}^n} (1 + |u|^2)^{k/2} \partial^{\alpha+\beta} \phi(u)\, du$$

$$\le C(1 + |x|^2)^{k/2} \gamma_{m',k'}(\phi),\ \forall\, \beta \in \underline{N}^n.$$

where m' and k' are suitable integers and $\gamma_{m',k'}$ is one of the seminorms defining the topology of $\mathcal{S}$. If $\ell > k + n$ the last inequality implies

$$(1 + |x|^2)^{-\ell/2} \left|\frac{\partial^\beta F}{\partial x^\beta}(x)\right| \le C\cdot\gamma_{m',k'}\cdot(1 + |x|^2)^{(k-\ell)/2}; \qquad (6.9)$$

hence the left-hand side of (6.9) is integrable, $\forall\, \beta \in \underline{N}^n$. Therefore, $(1 + |x|^2)^{-\ell/2} F(x) \in \mathcal{D}_{L^1}$. The last part of the theorem follows immediately from (6.9). Q.E.D.

Suppose now that we are given $T \in \mathcal{O}'_C$ and $S \in \mathcal{S}'$. By the corollary of Theorem 6.5, we can write $S = \partial^\alpha\left[(1 + r^2)^{k/2} f(x)\right]$, where f(x) is a bounded continuous function on $\underline{R}^n$. On the other hand, by Definition 6.3, $(1 + |x|^2)^{\ell/2}\, T \in \mathcal{D}'_{L^\infty}$, where $\ell > k + n$. Let us set

$$\langle T_\xi, \langle S_\eta, \phi(\xi + \eta)\rangle\rangle = \langle (1 + |\xi|^2)^{\ell/2}\, T_\xi, (1 + |\xi|^2)^{-\ell/2} F(\xi)\rangle \qquad (6.10)$$

and immediately observe that if we take, in particular, $T \in E' \subset \mathcal{O}_C'$ and $\phi \in C_c \subset \mathcal{S}$, the left-hand side of (6.10) coincides with $\langle S*T, \phi\rangle$. By the lemma 6.2, the right-hand side of (6.10) is well defined for all $T \in \mathcal{O}_C'$, $S \in \mathcal{S}'$, and $\phi \in \mathcal{S}$ and it can be shown that it depends continuously on each variable when the other two remain bounded [28, Chapter VII, p. 103]. The precise proof would require properties of the strong topology of $\mathcal{O}_C'$ which we have not defined here. However, we can give an idea of the proof by using the notion of convergence in the three different spaces involved. Indeed, if $\phi_j \to 0$ in $\mathcal{S}$ then, by Lemma 6.2, $F_j \to 0$ in $\mathcal{D}_{L^1}$; hence, the right-hand side of (6.10) converges to zero. On the other hand, if $T_j \to 0$ in $\mathcal{O}_C'$ then, by the definition of convergence in this space, $(1 + |\xi|^2)^{k/2} T_j \to 0$ in $\mathcal{D}_{L^\infty}'$ and, again by Lemma 6.2, the right-hand side of (6.10) converges to zero. Finally, if $S_j \to 0$ in $\mathcal{S}'$ we also get that (6.10) converges to zero. We have then sketched the proof of the following theorem.

Theorem 6.7. *If* $T \in \mathcal{O}_C'$ *and* $S \in \mathcal{S}'$, *then* $S*T$ *is well defined by* (6.10) *and belongs to* $\mathcal{S}'$. *Moreover, the map*

$$\mathcal{S}' \times \mathcal{O}_C' \ni (S,T) \to S*T \in \mathcal{S}'$$

is separately continuous.

As we know, $\mathcal{O}_M$ and $\mathcal{O}_C'$ are subspaces of $\mathcal{S}'$; hence the Fourier transform $\mathcal{F}$ is well defined. We shall see that $\mathcal{F}$ interchange these two spaces. More precisely, we have the following theorem.

Theorem 6.8. *The Fourier transform* $\mathcal{F}$ *is a one-to-one map from* $\mathcal{O}_M$ *onto* $\mathcal{O}_C'$. *Similarly* $\mathcal{F}$ *is a one-to-one map from* $\mathcal{O}_C'$ *onto* $\mathcal{O}_M$.

Proof. Since the inverse Fourier transform $\mathcal{F}^{-1}$ has the same properties as $\mathcal{F}$, it suffices to show that $\mathcal{F}$ maps $\mathcal{O}_M$ into $\mathcal{O}_C'$ and $\mathcal{O}_C'$ into $\mathcal{O}_M$.

Let $\phi \in \mathcal{O}_M$ and let m be an integer such that $m > n/2$. By Definition 4.9, for every integer $k \geq 0$ there is an integer ℓ

sufficiently large that

$$|(1 - \Delta)^k \phi(x)| \leq C(1 + r^2)^{\ell - m}, \forall\, x \in \underline{R}^n.$$

Setting

$$h(x) = \frac{(1 - \Delta)^k \phi(x)}{(1 + r^2)^{\ell}},$$

it is clear by the choice of the integer m that h is an integrable function. By taking Fourier transforms we get

$$(1 + |\xi|^2)^k\, \hat{\phi}(\xi) = (1 - \Delta_{\xi})^{\ell}\, \hat{h}(\xi)$$

where the right-hand side is a finite sum of derivatives of bounded functions. Therefore, for every integer $k \geq 0$, $(1 + |\xi|^2)^k\, \hat{\phi}(\xi) \in \mathcal{D}'_{L^\infty}$, which proves that $\hat{\phi} \in \mathcal{O}'_C$.

If $T \in \mathcal{O}'_C$, it follows from Definition 6.3 and Corollary 2 of Theorem 6.6 that for all $\alpha \in \underline{N}^n$, $x^\alpha T \in \mathcal{D}'_L$. By Theorem 6.3, we get that

$$x^\alpha T = D_\xi^\alpha\, \hat{T}(\xi)$$

is a continuous function slowly increasing at infinity, $\forall \alpha \in \underline{N}^n$. Hence, $\hat{T} \in \mathcal{O}_M$. Q.E.D.

Remark. It is also true that F is a topological isomorphism from $\mathcal{O}_M$ onto $\mathcal{O}'_C$ and from $\mathcal{O}'_C$ onto $\mathcal{O}_M$. (See Schwartz [28, Chapter VII, p. 125].)

As an application of the preceding theorem we shall prove the following.

Theorem 6.9. *If* $T \in \mathcal{O}'_C$ *and* $\alpha \in \mathcal{S}$ *then* $T*\alpha \in \mathcal{S}$.

Proof. It suffices to show that the Fourier transform $T*\alpha \in \mathcal{S}$. Since $\mathcal{O}'_C \subset \mathcal{S}'$ then by Theorem 4.9, $T*\alpha \in \mathcal{O}_M \subset \mathcal{S}'$; thus $T*\alpha$ is well defined and by Theorem 4.11, $T*\alpha = \hat{T}\cdot\hat{\alpha}$. But, by Theorem 6.8, $\hat{T} \in \mathcal{O}_M$ and by Proposition 4.3, $\hat{T}\cdot\hat{\alpha} \in \mathcal{S}$. Q.E.D.

We can now generalize the results of Theorems 4.11 and 4.13.

Theorem 6.10. *If* $T \in \mathcal{O}'_C$ *and* $S \in \mathcal{S}'$ *then*

$$S*T = \hat{S}\cdot\hat{T}. \tag{6.11}$$

Proof. By the corollary of Theorem 6.5 we can write

$$S = D^{\alpha}\Phi$$

where Φ is a continuous function slowly increasing at infinity. For all $\phi \in \mathcal{S}$ we have, taking into account (6.10),

$$\langle S*T, \phi\rangle = \langle S*T, \hat{\phi}\rangle = \langle T_\xi, \langle S_\eta, \hat{\phi}(\xi + \eta)\rangle\rangle.$$

On the other hand,

$$\begin{aligned} \langle S_\eta, \hat{\phi}(\xi + \eta) &= (-1)^{|\alpha|} \int_{\underline{R}^n} \Phi(\eta) D^{\alpha}_{\eta}\hat{\phi}(\xi + \eta)\, d\eta \\ &= (-1)^{|\alpha|} \int_{\underline{R}^n} \Phi(\eta - \xi) D^{\alpha}_{\eta}\hat{\phi}(\eta)\, d\eta = \int_{\underline{R}^n} \Phi(\eta - \xi)\widehat{y^{\alpha}\phi}(\eta)\, d\eta \\ &= (\check{\Phi}*\widehat{y^{\alpha}\phi})(\xi). \end{aligned}$$

As in the proof of Theorem 4.11, it is easy to check that

$$\check{\Phi} = (2\pi)^{-n}\hat{\hat{\Phi}}$$

and that

$$\widehat{\Phi*y^{\alpha}\phi} = (2\pi)^{-n}\hat{\hat{\Phi}}*\widehat{y^{\alpha}\phi} = \widehat{\hat{\Phi}\cdot y^{\alpha}\phi}$$

(see Problem 18 of Chapter 4). Making the appropriate replacement

$$<S*T, \phi> = <T_\xi, <S_\eta, \hat{\phi}(\xi+\eta)>> = <T_\xi, \widehat{\Phi\cdot y^\alpha\phi}(\xi)>$$

$$= <\hat{T}, \hat{\Phi}\cdot y^\alpha\phi> = <\hat{T}, \widehat{D_\eta^\alpha\Phi\cdot\phi}> = <\hat{T}, \hat{S}\cdot\phi> = <\hat{S}\cdot\hat{T}, \phi>$$

because $\hat{T} \in \mathcal{O}_M$ and $\hat{S} \in \mathcal{S}'$. Q.E.D.

In a similar way we can extend to distributions Property IV of Chapter 4, Section 3 by the following theorem.

Theorem 6.11. *If* $T \in \mathcal{O}_M$ *and* $S \in \mathcal{S}'$, *then*

$$S\cdot T = (2\pi)^{-n}\hat{S}*\hat{T}. \tag{6.12}$$

PROBLEMS

1. Prove that the topology of $\mathcal{D}_{L^p}$ coincides with the coarsest one for which the imbeddings $\mathcal{D}_{L^p} \to H^{m,p}$ are continuous, $\forall m$.

2. If $1 < p < +\infty$, prove that $\mathcal{D}_{L^p}$ is a reflexive space.

3. Complete the proof of Theorem 6.2 in the case $p = 1$.

4. Prove that $T \in \mathcal{O}_C'$ if and only if for all polynomials $P(x)$ $PT \in \mathcal{D}_{L^\infty}'$.

5. Prove that every element of $\mathcal{O}_C'$ defines a tempered distribution.

6. Prove Corollary 1 of Theorem 6.6.

7. Complete the proof of Lemma 6.2 by proving that if $\phi_j \to 0$ in $\mathcal{S}$ then $F_j \to 0$ in $\mathcal{D}_{L^1}$.

8. Show that if $T \in \mathcal{O}_C'$ then $x^\alpha T \in \mathcal{D}_{L^1}'$, $\forall\alpha \in N^n$.

9. Prove that: (i) if $h \in \mathcal{D}_{L^\infty}$ and $\phi \in \mathcal{D}_{L^1}$ then $h\phi \in \mathcal{D}_{L^1}$; (ii) if $h \in \mathcal{D}_{L^\infty}$ and $T \in \mathcal{D}_{L^\infty}'$ then $hT \in \mathcal{D}_{L^\infty}'$.

10. If $\phi \in \mathcal{D}_{L^\infty}$ and $f \in H^s$ with s an integer, prove that $\phi\, f \in H^s$.

Chapter 7

APPLICATIONS

1. LOCAL AND PSEUDOLOCAL OPERATORS

If $P(x, D) = \sum_{|p|\le m} a_p(x)D^p$ is a partial differential operator with C^∞ coefficients, it is obvious that

$$\operatorname{supp} PT \subset \operatorname{supp} T, \ \forall T \in \mathcal{D}'.$$

In other words, a partial differential operator shrinks the support of a distribution. Such a property motivates the following definition.

Definition 7.1. *A continuous linear operator* L *from* $C_c^\infty(\Omega)$ *into* $C^\infty(\Omega)$ *is said to be a local operator if it can be extended to a continuous linear operator from* $\mathcal{E}'(\Omega)$ *into* $\mathcal{D}'(\Omega)$ *and it is such that*

$$\operatorname{supp} LT \subset \operatorname{supp} T, \ \forall T \in \mathcal{E}'(\Omega).$$

Thus every partial differential operator with C^∞ coefficients is a local operator. Conversely, Peetre [23, 24] has shown that every operator of local type is a partial differential operator.

In order to define pseudolocal operators we have to introduce the notion of *singular support* of a distribution.

Definition 7.2. *We call singular support of a distribution* T *the complement of the largest open set where* T *is a* C^∞ *function.*

Obviously

$$\text{sing supp } T \subset \text{supp } T$$

because in the complement of its support, T coincides with the function identically zero.

Examples. 1. The singular support of any function in $C^\infty(\underline{R}^n)$ is the empty set.

2. The singular support of the Dirac measure δ is the origin. In this case, the singular support coincides with the support.

3. The support of the characteristic function of an open interval (a,b) is the closed interval [a,b]. Its singular support is the set $\{a\} \cup \{b\}$.

Definition 7.2. *A continuous linear operator* L *from* $C_c^\infty(\Omega)$ *into* $C^\infty(\Omega)$ *is said to be a pseudolocal operator if it satisfies the following conditions:* (i) L *can be extended to a continuous linear operator from* $\mathcal{E}'(\Omega)$ *into* $\mathcal{D}'(\Omega)$; (ii) *for every* $T \in \mathcal{E}'(\Omega)$ *we have*

$$\text{sing supp } LT \subset \text{sing supp } T.$$

Examples. 1. Let $\phi \in C^\infty(\underline{R}^n)$. The operator M_ϕ defined by

$$M_\phi(\psi) = \phi\cdot\psi, \ \forall\psi \in C_c^\infty(\underline{R}^n),$$

is pseudolocal. This operator is obviously a local one.

2. Let $\phi \in C^\infty(\underline{R}^n)$. The operator $L_\phi(\psi) = \phi*\psi$ is pseudolocal. Indeed, by Theorem 3.2,

$$\text{sing supp } (\phi*T) = \emptyset, \ \forall T \in \mathcal{E}'(\underline{R}^n).$$

The following result provides a nontrivial example of a pseudolocal operator. (See Schwartz [29].

Theorem 7.1. *Let* $E \in \mathcal{D}'(\underline{R}^n)$ *and suppose that* E *is a* C^∞ *function in* $\underline{R}^n$-{0}, *the complement of the origin in* $\underline{R}^n$. *The operator*

L_E *defined by*

$$L_E(\phi) = E*\phi, \ \forall\phi \in C_c^\infty(\underline{R}^n),$$

is pseudolocal.

In order to prove Theorem 7.1, it suffices to prove Condition (ii) of Definition 7.2 since, by Property 2 of Chapter 3, Section 2, the convolution product extends continuously to $\mathcal{E}'(\underline{R}^n)$. To prove Condition (ii) of Definition 7.2, it suffices to show, by taking complements, that *if* $T \in \mathcal{E}'(\underline{R}^n)$ *and* T *is a* C^∞ *function on an open set* $\Omega \subset \underline{R}^n$ *then* $E*T$ *is a* C^∞ *function on* Ω. This result will follow from a slightly more general one about regularization of distributions (see Chapter 3, Section 3). First, let us introduce the following notations.

If Ω is an open set in $\underline{R}^n$ and $y \in \underline{R}^n$, let

$$\Omega - y = \{x - y: x \in \Omega\}$$

be the *translation of* Ω *by* y. Clearly, $\Omega - y$ is an open set.

If A is any subset of $\underline{R}^n$, let

$$\Omega - A = \bigcup_{y \in A} \Omega - y.$$

Clearly, $\Omega - A$ is an open set. If $A \subset B$, we have $\Omega - A \subset \Omega - B$.

For every real number $\varepsilon > 0$ let

$$\Omega^\varepsilon = \{x \in \Omega: \ d(x, \Omega^c) > \varepsilon\},$$

where Ω^c denotes the complement of Ω in $\underline{R}^n$. The set Ω^ε is also an open set and we have $\Omega = \Omega^\varepsilon + B_\varepsilon$, where B_ε denotes the ball with center at the origin and radius ε.

Lemma 7.1. *Let* $T \in \mathcal{E}'(\underline{R}^n)$, *let* K *be the support of* T, *and let* Ω *be any open subset in* $\underline{R}^n$. *If a distribution* $E \in \mathcal{D}'(\underline{R}^n)$ *is zero* *(see Definition 2.4) on* Ω - K*, then the convolution product* E*T *is zero on* Ω.

Proof. By assumption, we have

$$\operatorname{supp} T = K \text{ and } \operatorname{supp} E \subset (\Omega - K)^c.$$

By Property 1 of Chapter 3, Section 2, we have

$$\operatorname{supp}(E*T) \subset \overline{K + (\Omega - K)^c}.$$

But, for all $x \in K$ and all $y \in (\Omega - K)^c$ we must have $x + y \in \Omega^c$, which implies that

$$\overline{K + (\Omega - K)^c} \subset \Omega^c;$$

hence E*T is zero on Ω. Q.E.D.

Corollary. Under the assumptions as for the lemma, suppose that the distribution E *is zero on* $(\Omega - K)^\varepsilon$. *Then the convolution product* E*T *is zero on* Ω^ε.

Proof. In the proof of Lemma 7.1 we have obtained the following inclusion: $K + (\Omega - K)^c \subset \Omega^c$. On the other hand, $(\Omega^\varepsilon)^c = \Omega^c + B_\varepsilon$; hence

$$K + (\Omega - K)^c + B_\varepsilon \subset \Omega^c + B_\varepsilon = (\Omega^\varepsilon)^c$$

which implies

$$\overline{K + (\Omega - K)^c} \subset (\Omega^\varepsilon)^c;$$

Therefore, E*T is zero on Ω^ε. Q.E.D.

Lemma 7.2. *Under the same assumptions as for Lemma* 7.1, *suppose that the distribution* E *is a* C^∞ *function on* Ω - K. *Then,* E*T *is a* C^∞ *function on* Ω.

Proof. Since differentiability is a local property we can assume that Ω is a *bounded* set. Now let β be a C^∞ function equal to one on $(\Omega - K)^\varepsilon$ and having its support contained in $(\Omega - K)^{\varepsilon/2}$ Set $E_1 = \beta E$. From our assumptions, it follows that E_1 is a C^∞ function with compact support in $\underline{R}^n$; hence the convolution product $E_1 * T$ is a C^∞ function in $\underline{R}^n$. On the other hand, it follows from our choice of the function β that E_1 coincides with E on $(\Omega - K)^\varepsilon$. Hence, by the corollary of Lemma 7.1, $E_1 * T$ coincides with $E*T$ on Ω^ε, that is, $E*T$ is a C^∞ function on Ω^ε. By letting $\varepsilon \to 0$, we get that $E*T$ is C^∞ on Ω. Q.E.D.

Proof of Theorem 7.1. As we remarked above, it suffices to show that if $T \in \mathcal{E}'(\underline{R}^n)$ and T is a C^∞ function on an open set $\Omega \subset \underline{R}^n$, the convolution $E*T$ is a C^∞ function on Ω. Let ω be a relatively compact open set such that $\bar{\omega} \subset \Omega$ and let $\alpha \in C_c^\infty(\Omega)$ be such that $\alpha = 1$ on a neighborhood of $\bar{\omega}$. Write

$$E*T = E*\alpha T + E*(1 - \alpha)T.$$

Since $\alpha T \in C_c^\infty(\underline{R}^n)$, it follows that $E*\alpha T \in C^\infty(\underline{R}^n)$ (Theorem 3.2). On the other hand, if K denotes the support of $(1 - \alpha)T$ then, by our choice of the function α, we have $\bar{\omega} \cap K = \emptyset$; hence $\omega - K \subset \underline{R}^n - \{0\}$. But, by assumption, E is a C^∞ function on $\underline{R}^n - \{0\}$, *a fortiori* on $\omega - K$; hence, by Lemma 7.2, $E*(1 - \alpha)T$ is a C^∞ function on ω. Therefore, $E*T$ is a C^∞ function on ω and, since ω is an arbitrary relatively compact open subset of Ω, $E*T$ is a C^∞ function on Ω. Q.E.D.

2. HYPOELLIPTIC PARTIAL DIFFERENTIAL OPERATORS

The example of a pseudolocal operator given by Theorem 7.1 is related to the theory of partial differential operators.

Definition 7.3. *Let*

$$P = P(x, D) = \sum_{|p| \leq m} a_p(x) D^p$$

be a partial differential operator with C^∞ coefficients on Ω. We say that P is hypoelliptic in Ω if it satisfies the following condition:

(H) for every open subset U of Ω and every distribution T in Ω, the fact that $PT \in C^\infty(U)$ implies $T \in C^\infty(U)$.

In other words, all solutions u of the equation $Pu = f$ belong to $C^\infty(U)$ whenever $f \in C^\infty(U)$.

From now on we shall only consider partial differential operators with *constant coefficients*. It follows from Definition 7.3 that if the operator P is hypoelliptic and has a fundamental solution (see Definition 2.8) then E is a C^∞ function off the origin. Conversely, we have the following result.

Theorem 7.2. *Let* P *be a partial differential operator with constant coefficients and suppose that* P *has a fundamental solution* E *which is a* C^∞ *function on* $\underline{R}^n - \{0\}$. *Then* P *is hypoelliptic in* $\underline{R}^n$.

Proof. We have to show that Condition (H) is satisfied. Let $T \in \mathcal{D}'(\underline{R}^n)$ and suppose that the distribution $S = PT$ is a C^∞ function on some open set Ω. Let $\alpha \in C_c^\infty(B_\varepsilon)$, where B_ε is a ball with center at the origin and radius ε be such that α is equal to one on a neighborhood of the origin and set $F = \alpha E$. By Leibniz's rule, we have

$$PF = \sum_{|p| \leq m} a_p D^p(\alpha E) = \alpha PE + \sum_{|r|>0} a_p c_{rs} D^r\alpha \cdot D^s E = \delta + \beta$$

where $\beta \in C_c^\infty(B_\varepsilon)$. The distribution F is said to be a parametrix of P.

Next write

$$T = \delta * T = PF * T - \beta * T = F * PT - \beta * T = F * S - \beta * T.$$

Since $\beta \in C_c^\infty(B)$, the convolution $\beta * T$ belongs to $C^\infty(\underline{R}^n)$. Let $K = \operatorname{supp} F \subset B$. We have $\Omega^\varepsilon - K \subset \Omega^\varepsilon - B_\varepsilon \subset \Omega$. By assumption, S is a C^∞ function on Ω, *a fortiori* on $\Omega^\varepsilon - K$; hence, by Lemma 7.2, $F*S$ is a C^∞ function on Ω^ε. By letting $\varepsilon \to 0$, we conclude that T is a C^∞ function on Ω, Q.E.D.

In the next section, we shall prove that every partial differential operator with constant coefficients has a fundamental solution. Combining such a result with Theorem 7.2 and the definition of hypoellipticity we get the following characterization of hypoelliptic partial differential operators.

A partial differential operator with constant coefficients is hypoelliptic if and only if it has a fundamental solution which is a C^∞ *function off the origin.*

3. EXISTENCE OF FUNDAMENTAL SOLUTIONS

In Chapter 2, Section 5 we defined, with a few simple examples, the notion of fundamental solution of a partial differential operator.

Fundamental solutions are very useful tools in the theory of partial differential equations, for instance, in solving inhomogeneous equations, in providing information about the regularity and growth of solutions. Also, as we remarked at the end of the last section, certain partial differential operators can be characterized by properties of their fundamental solutions.

If E is a fundamental solution of $P = P(D)$ and if f is distribution, then, taking into account (3.13), we see that a solution of the inhomogeneous equation

$$Pu = f$$

is given by $u = E*f$ whenever the convolution product is defined.

Fundamental solutions of classical differential operators like the *Laplace operator*, the *heat operator*, and the *Cauchy-*

Riemann operator were known for a long time. The conjecture that every partial differential operator with constant coefficients has a fundamental solution was proved in 1954 by Malgrange [21] (see also [29]) and independently by Ehrenpreis [8]. After that, several other proofs, obtaining more precise information about the fundamental solutions, have appeared in the literature. The reader should consult the books of Hörmander [14] and Treves [31]. We should also mention the article of Hörmander [16], in which, by solving a *division problem* conjectured by Schwartz, he proves the existence of a *tempered* fundamental solution for every partial differential operator with constant coefficients.

Before reproducing Malgrange's original proof we shall give some classical examples of fundamental solutions and we shall present a method of constructing fundamental solutions of *homogeneous elliptic operators*.

Laplace Operator.

Let

$$\Delta = \sum_{j=1}^{n} \frac{\partial^2}{\partial x_j^2}$$

be the Laplace operator in the space $\underline{R}^n$. If $n > 2$ the distribution

$$E = \frac{\Gamma(n/2)}{(2-n)2\pi^{n/2}} r^{2-n}$$

is a fundamental solution of Δ. If $n = 2$ the distribution

$$E = \frac{1}{2\pi} \log r$$

is a fundamental solution of Δ.

Let us prove this in the case $n > 2$. For all $\phi \in C_c^\infty(\underline{R}^n)$ we have

$$<\Delta(r^{2-n}),\phi> = <r^{2-n},\Delta\phi> = \lim_{\varepsilon\to 0} \int_{r\geq\varepsilon} r^{2-n}\Delta\phi \, dx.$$

By applying Green's formula[1] and remarking that $\Delta(r^{2-n}) = 0$ when $r \geq \varepsilon$, we get

$$<\Delta(r^{2-n}),\phi> = \lim_{\varepsilon\to 0} \left\{ \iint_{r=\varepsilon} \frac{d}{dr}(r^{2-n})\phi \, d\omega_\varepsilon - \int_{r=\varepsilon} r^{2-n} \frac{d\phi}{dr} \, d\omega_\varepsilon \right\},$$

where $d\omega_\varepsilon$ denotes the area element of the sphere with center at the origin and radius ε. Since $d\phi/dr$ is bounded near the origin and the area of the sphere of radius ε is equal to $A\cdot\varepsilon^{n-1}$, where A is the area of the unit sphere, the last integral is bounded by a constant times ε, hence it tends to zero as $\varepsilon \to 0$. On the other hand, the first integral can be written as follows:

$$\int_{r=\varepsilon} \frac{d}{dr}(r^{2-n})\phi \, d\omega_\varepsilon = (2-n)\int_{r=\varepsilon} \varepsilon^{1-n}\phi \, d\omega_\varepsilon = (2-n)\int_{r=\varepsilon} \phi d\omega,$$

where $d\omega$ is the area element of the unit sphere. When $\varepsilon \to 0$, the last integral converges to $A\cdot\phi(0)$, where

$$A = \frac{2\pi^{n/2}}{\Gamma(n/2)}$$

is the area of the unit sphere. Making the appropriate replacement, we get

$$<\Delta(r^{2-n}),\phi> = \frac{(2-n)2\pi^{n/2}}{\Gamma(n/2)}\phi(0), \ \forall\phi \in C_c^\infty(\underline{R}^n);$$

---

[1] $\int_V (u\,\Delta v - \Delta\, uv)\, dV = \int_S [u(\partial v/\partial\eta) - v(\partial u/\partial\eta)]\, dS$, where V is a volume, S its boundary, η the inner normal, dV the volume element, and dS the area element.

hence E is a fundamental solution of Δ.

Cauchy-Riemann Operator

Let $z = x + iy$ be a complex variable and let

$$\frac{\partial}{\partial \bar{z}} = \frac{1}{2}\left(\frac{\partial}{\partial x} + i\frac{\partial}{\partial y}\right)$$

be the Cauchy-Riemann operator. It can be shown that

$$E = \frac{1}{\pi z}$$

is a fundamental solution of the Cauchy-Riemann operator [31, p. 242].

Heat Operator

In the (n + 1)-dimensional Euclidean space $\underline{R}^{n+1}$, denote by $(x, t) = (x_1,\cdots,x_n,t)$ a variable element, by $x = (x_1,\cdots,x_n)$ the space variable, and by t the time variable. The partial differential operator

$$H = \frac{\partial}{\partial t} - \Delta$$

where Δ is the Laplace operator in the space variables, is called the *heat operator*. The distribution

$$E(x, t) = \left(\frac{1}{2\sqrt{\pi t}}\right)^n Y(t) \exp - \frac{|x|^2}{4t}$$

is a fundamental solution of H [31, p. 245].

Elliptic Operators

The Laplace operator Δ is the classical example of an elliptic partial differential operator, whose definition goes as follows. Let

$$P = P(D) = \sum_{|p|\le m} a_p D^p$$

be a differential operator of order m with constant coefficients. The homogeneous polynomial of degree m

$$P_m(\xi) = \sum_{|p|=m} a_p \xi^p$$

is said to be the *characteristic polynomial* of P(D).

Definition 7.4. *We say that the partial differential operator* P(D) *is elliptic if the following condition is satisfied:*

(E) $P_m(\xi) \neq 0$ *for all* $\xi \in \underline{R}^n$ *such that* $\xi \neq 0$.

This condition is easily seen to be equivalent to the following one.

(E') *There is a constant* $c_0 > 0$ *(ellipticity constant) such that*

$$|P_m(\xi)| \geq c_0 |\xi|^m \ \forall \xi \in \underline{R}^n.$$

The Laplace operator Δ, powers of the Laplace operator Δ^m, and the Cauchy-Riemann operator $\partial/\partial\bar{z}$ are examples of elliptic operators.

Formally, the problem of constructing a fundamental solution of a partial differential operator P = P(D) is a very easy one. Indeed, suppose that we have

$$P(D)E = \delta.$$

Then, by taking Fourier transforms we obtain

$$P(\xi)\cdot\hat{E} = 1;$$

hence

$$\hat{E} = \frac{1}{P(\xi)}$$

and E should be defined as the inverse Fourier transform of $1/P(\xi)$. However, this does not always make sense, because of the *real zeros* of the polynomial $P(\xi)$. This difficulty can be overcome, for instance, by selecting a domain of integration in $\underline{C}^n$ which avoid the zeros of $P(\xi)$. We shall not discuss this method here and the reader should consult the books by Hormander [14] and Treves [31].

Suppose that $P = P_m$ is a homogeneous elliptic partial differential operator of degree m with constant coefficients. In order to construct its fundamental solution, we shall prove that the homogeneous function of degree -m

$$U(\xi) = \frac{1}{P_m(\xi)}$$

defines a tempered distribution whose inverse Fourier transform is the fundamental solution of the operator P.

Definition 7.5. *A function* $U(x)$ *defined on* $\underline{R}^n$ *is said to be homogeneous of degree* λ, *a complex number, if*

$$U(tx) = t^{\lambda}U(x), \ \forall t > 0.$$

We can always write $U(x) = r^{\lambda}f(\omega)$, where $r = |x|$ and $\omega = (\omega_1, \cdots, \omega_n) \in \sum$, the unit sphere in $\underline{R}^n$. For simplicity, we are going to consider only the case when λ is an integer. For the general case, the reader should consult Refs. [9, 28]. Also, we shall assume that U is a C^{∞} function off the origin. Thus, let

$$U(x) = r^{-m}f(\omega) \tag{7.1}$$

where m is an integer $\gtreqless 0$ and f is a C^{∞} function on $\sum$. We claim that the function U defines a *tempered distribution* on $\underline{R}^n$. To prove this we have to consider two cases, namely $m < n$ and $m \geq n$.

Suppose $m < n$. Define for every $\phi \in S(\underline{R}^n)$

$$u_{\phi}(r) = \int_{\Sigma} f(\omega)\phi(r\omega)\ d\omega. \tag{7.2}$$

It is easy to see that $u_\phi(r) \in S(\underline{R})$ for all $\phi \in S(\underline{R}^n)$. Next define

$$\langle U,\phi\rangle = \int_0^\infty r^{-m+n-1} u_\phi(r)\, dr, \ \forall\phi \in S(\underline{R}^n). \tag{7.3}$$

Since $\phi \in S$ and $m < n$, the integral (7.3) is absolutely convergent. Also, (7.3) depends continuously on $\phi \in S$; therefore, U defines a tempered distribution.

Suppose $m \geq n$. In this case the integral (7.3) is not well defined. We shall then use the notion of *finite part* of a divergent integral in order to define U as a tempered distribution.

Suppose, first, that $m = n$ and consider the integral

$$\int_\varepsilon^\infty r^{-1} u_\phi(r)\, dr = \int_\varepsilon^1 r^{-1} u_\phi(r)\, dr + \int_1^\infty r^{-1} u_\phi(r)\, dr. \tag{7.4}$$

The second integral on the right-hand side of (7.4) is convergent. Let us write the first integral on the right-hand side of (7.4) as follows:

$$\begin{aligned}\int_\varepsilon^1 r^{-1} u_\phi(r)\, dr &= \int_\varepsilon^1 r^{-1}(u_\phi(r) - u_\phi(0))\, dr + \int_\varepsilon^1 u_\phi(0)\cdot r^{-1} dr \\ &= \int_\varepsilon^1 r^{-1}(u_\phi(r) - u_\phi(0))\, dr - u_\phi(0)\cdot \log \varepsilon.\end{aligned}$$

Making replacements in (7.4) we get

$$\int_\varepsilon^\infty r^{-1} u_\phi(r)\, dr = \int_\varepsilon^1 r^{-1}(u_\phi(r) - u_\phi(0))dr + \int_1^\infty r^{-1} u_\phi(r)\, dr - u_\phi(0)\cdot \log \varepsilon. \tag{7.5}$$

By writing

$$F(\varepsilon) = \int_\varepsilon^1 r^{-1}(u_\phi(r) - u_\phi(0))dr + \int_1^\infty r^{-1} u_\phi(r)\, dr$$

and

$$I(\varepsilon) = -u_\phi(0)\cdot \log \varepsilon,$$

we see that the integral (7.5) decomposes into two parts: $F(\varepsilon)$, which has a finite limit as $\varepsilon \to 0$, plus $I(\varepsilon)$, which tends to infinity as $\varepsilon \to 0$. By definition, we call the limit of $F(\varepsilon)$ as $\varepsilon \to 0$ the *finite part* of the integral (7.3) with $m = n$. We write

$$\mathrm{FP}\int_0^\infty r^{-1}u_\phi(r)\,dr = \int_0^1 r^{-1}(u_\phi(r) - u_\phi(0))\,dr + \int_1^\infty r^{-1}u_\phi(r)\,dr. \quad (7.6)$$

In a similar way, one can show that when $m > n$ the finite part of the integral (7.3) is equal to

$$\mathrm{FP}\int_0^\infty r^{-m+n-1}u_\phi(r)\,dr = \int_0^1 r^{-m+n-1}\left\{u_\phi(r) - \sum_{s=0}^{m=n} \frac{1}{s!}\frac{\partial^s u_\phi(0)}{\partial r^s} r^s\right\} dr$$
$$+ \int_0^\infty r^{-m+n-1}\left\{u_\phi(r) - \sum_{s=o}^{m-n-1} \frac{1}{s!}\frac{\partial^s u_\phi(0)}{\partial r^s} r^s\right\} dr. \quad (7.7)$$

It is obvious that when $m < n$ the finite part of (7.3) coincides with the value of that integral.

We observe that if $\phi_j \to 0$ in $S(\underline{R}^n)$ then $u_{\phi j} \to 0$ in $S(\underline{R}^n)$ and consequently the right-hand sides of (7.6) and (7.7) converge to zero. Summarizing all these results, we can say that if $U(x)$ is a homogeneous function of degree $-m$, C^∞ outside the origin, it defines a tempered distribution by

$$\langle U, \phi\rangle = \mathrm{FP}\int_0^\infty r^{-m+n-1}u_\phi(r)\,dr, \quad \phi \in S \quad (7.8)$$

When $m < n$ the symbol FP is not necessary.

Since $U \in S'$, its Fourier transform $\hat{U}$ is well defined, belongs to S', and has the following homogeneity properties: (i) if $m < n$, $\hat{U}$ is a C^∞ function outside the origin and homogeneous of degree $m - n$; (ii) if $m \geq n$, $\hat{U}$ decomposes into the sum

$$\hat{U} = H + P\cdot \ln|x|$$

where H is a homogeneous function of degree $m - n$ which is C^∞

outside the origin and P is a homogeneous polynomial of degree m - n. For the proof of these results, the reader should consult the article [1].

In conclusion, *if*

$$P(D) = \sum_{|p|=m} a_p D^p$$

is a homogeneous elliptic operator of order m *and has constant coefficients, then* P(D) *has a tempered fundamental solution* E *such that:* (i) *if* $m < n$, E *is a homogeneous function of degree* $m - n$, C^∞ *outside the origin;* (ii) if $m \geq n$, $E = H(x) + P(x)\cdot \ln|x|$, where H is a homogeneous *function of degree* $m - n$, C^∞ *outside the origin, and* P *a homogeneous polynomial of degree* $m - n$.

We now prove Malgrange's theorem on the existence of fundamental solutions.

Theorem 7.3. *Every partial differential operator with constant coefficients in* $\underline{R}^n$ *has a fundamental solution* E *belonging to* $\mathcal{D}'^{(n+1)}(\underline{R}^n)$, *the space of distributions of order* $n + 1$.

The proof of this theorem will be based on two lemmas.

Lemma 7.3. *Let* f(z) *be an analytic function of a complex variable on the unit disk* $\{z \in \underline{C}: |z| \leq 1\}$ *and let* p(z) *be a polynomial with leading coefficient* a. *Then, the following inequality holds true:*

$$|af(0)| \leq \frac{1}{2\pi}\int_0^{2\pi} |f(e^{i\theta})p(e^{i\theta})|\, d\theta. \tag{7.9}$$

Proof. Let m be the degree of p(z) and write

$$p(z) = az^m + bz^{m-1} + \cdots.$$

Let

$$\bar{p}(z) = \bar{a}z^m + \bar{b}z^{m-1} + \cdots$$

be the polynomial obtained by conjugating the coefficients of p. If we set

$$q(z) = z^m \bar{p}\left(\frac{1}{z}\right),$$

it is easy to see that

$$\bar{q}(0) = \bar{a} \text{ and } |p(e^{i\theta})| = |\bar{q}(e^{i\theta})|.$$

By Cauchy's formula, we have

$$f(0)\bar{q}(0) = \frac{1}{2\pi i} \int_{|z|=1} \frac{f(z)\bar{q}(z)}{z}\, dz$$

which immediately yields (7.9). Q.E.D.

Lemma 7.4. *Let* f(z) *be an entire function in* $\underline{C}$ *and let* p(z) *be a polynomial with leading coefficient* a. *Then for all* $z_0 \in C$ *we have the inequality*

$$|af(z_0)| \le \sup_{|z-z_0|\le 1} |f(z)p(z)|. \tag{7.10}$$

Proof. By applying Lemma 7.3 to $f(z_0 + z)$ and $p(z_0 + z)$ we get

$$|af(z_0)| \le \frac{1}{2\pi} \int_0^{2\pi} |f(z_0 + e^{i\theta})p(z_0 + e^{i\theta})|\, d\theta$$

which implies (7.10). Q.E.D.

Proof of Theorem 7.3. 1. Let $P = P(D) = \sum_{|p|\le m} a_p D^p$ be our partial differential operator with constant coefficients and let

$$^tP = {}^tP(D) = \sum_{|p| \le m} (-1)^{|p|} a_p D^p$$

be the transpose of P, which is defined by

$$\langle PT, \phi\rangle = \langle T, {}^tP\phi\rangle, \quad T \in \mathcal{D}', \ \forall\phi \in C_c^\infty. \tag{7.11}$$

Suppose that E is a fundamental solution of P. In (7.11), replacing T by E and taking into account that $PE = \delta$, we get

$$\langle E, {}^tP\phi\rangle = \phi(0), \ \forall\phi \in C_c^\infty. \tag{7.12}$$

Let ${}^tPC_c^\infty$ be the image of C_c^∞ by ${}^tP$. It is clear that ${}^tPC_c^\infty \subset C_c^\infty$. On the other hand, if $\psi \in {}^tPC_c^\infty$, there is a *unique* $\phi \in C_c^\infty$ such that $\psi = {}^tP\phi$. Indeed, if ${}^tP\phi = 0$ for some $\phi \in C_c^\infty$, by taking Fourier transform we get

$${}^tP(\xi)\cdot\hat{\phi}(\xi) = 0, \forall\xi \in \underline{R}^n,$$

which implies, since ${}^tP(\xi)$ is a polynomial, that $\hat{\phi}$ is identically zero.

Therefore, if E is a fundamental solution, the linear functional

$$E: \psi \in {}^tPC^\infty \to \langle E, \psi\rangle = \phi(0), \text{ where } {}^tP\phi = \psi,$$

it is well defined on ${}^tPC^\infty$ and it is continuous with respect to the topology induced on ${}^tPC_c^\infty$ by the natural topology of C_c^∞.

Conversely, if we can prove that the above linear functional is continuous on ${}^tPC_c^\infty$ *equipped with the topology induced by* C_c^k *for some* k, then by the Hahn-Banach theorem we can extend it to a continuous linear functional E on the whole space C_c^k. Hence, $E \in \mathcal{D}'^k$ and, by (7.12), $PE = \delta$, i.e., E is a fundamental solution of P.

2. Therefore, to prove the theorem, it suffices to show that *there is an integer* k, *sufficiently large, such that the linear*

functional

$$\psi \in {}^tPC_c^\infty \to \phi(0) \in \underline{C}, \text{ where } {}^tP\phi = \psi,$$

is continuous with respect to the topology of C_c^k.

By changing variables, if necessary, we can assume that the operator ${}^tP(D)$ can be written as follows:

$${}^tP(D) = D_n^m + \sum_{h=1}^{m} P_h(D_1,\cdots,D_{n-1})D_n^{m-h}$$

where P_h is a differential operator on $D_1,\cdots,D_{n-1}$. For simplicity let us use the following notations. The elements of the space $\underline{R}^n$ will be denoted by $x = (x', t)$, with $x' = (x_1,\cdots,x_{n-1})$ and $t = x_n$. We also set $D = (D_{x'}, D_t)$, with $D_{x'} = (D_1,\cdots D_{n-1})$ and $D_t = D_{x_n}$. The elements of the space $\underline{C}^n$ will be denoted by $\zeta = (\zeta', \tau)$, with $\zeta' = (\zeta_1,\cdots,\zeta_{n-1})$, $\zeta_j = \xi_j + i\eta_j$, $1 \leq j \leq n-1$, and $\tau = \mu + i\sigma$. With these notations, we can write

$${}^tP(D) = D_t^m + \sum_{h=1}^{m} P_h(D_{x'})D_t^{m-h}.$$

If $\phi \in C_c^\infty$, let $\hat{\phi}(\zeta)$ denote its Fourier-Laplace transform. By the Paley-Wiener theorem (Theorem 4.12), $\hat{\phi}(\zeta)$ is an entire function on $\underline{C}^n$ and such that $\hat{\phi}(\xi) \in S(\underline{R}^n)$. From Fourier's inversion formula (4.5), we easily get the inequality

$$|\phi(0)| \leq (2\pi)^{-n} \int_{\underline{R}^n} |\hat{\phi}(\xi', \mu)d\xi' \, d\mu. \tag{7.13}$$

We can estimate the last integral as follows:

$$\int_{\underline{R}^n} |\hat{\phi}(\xi', \mu)|d\xi' \, d\mu \leq$$

$$A \cdot \int_{\underline{R}^n} \frac{d\xi'\, d\mu}{1+|\xi_1|^{n+1}+\cdots+|\xi_{n-1}|^{n+1}+|\mu|^{n+1}} \cdot$$

where

$$A = \sup_{\xi\in\underline{R}^n} |(1+|\xi_1|^{n+1}+\cdots+|\xi_{n-1}|^{n+1}+|\mu|^{n+1})\hat{\phi}(\xi)|$$

By setting

$$M = (2\pi)^{-n} \int_{\underline{R}^n} \frac{d\xi'\, d\mu}{1+|\xi_1|^{n+1}+\cdots+|\xi_{n-1}|^{n+1}+|\mu|^{n+1}}$$

we get

$$|\phi(0)| \le A\cdot M. \tag{7.14}$$

3. In order to estimate A, we shall apply Lemma 7.4 to the following entire functions of a complex variable $\tau = \mu + i\sigma$:

$$\hat{\phi}(\xi', \tau),\ \xi_j^{n+1}\ \hat{\phi}(\xi', \tau),\ 1 \le j \le n-1, \text{ and } \tau^{n+1}\hat{\phi}(\xi', \tau)$$

and the polynomial ${}^tP(\xi', \tau)$, at the point μ. (Here, ξ' is assumed to be a fixed element in $\underline{R}^{n-1}$.) We get the following inequalities:

$$|\hat{\phi}(\xi',\mu)| \le \sup_{|\tau-\mu|\le 1} |{}^tP(\xi', \tau)\ \hat{\phi}(\xi', \tau)| \tag{7.15}$$

$$|\xi_j^{n+1}\ \hat{\phi}(\xi', \mu)| \le \sup_{|\tau-\mu|\le 1} |\xi_j^{n+1}\ {}^tP(\xi', \tau)\hat{\phi}(\xi', \tau)|,\ 1 \le j \le n-1, \tag{7.16}$$

$$|\mu^{n+1}\hat{\phi}(\xi', \mu)| \le \sup_{|\tau-\mu|\le 1} |\tau^{n+1}\ {}^tP(\xi', \tau)\hat{\phi}(\xi', \tau)|. \tag{7.17}$$

On the other hand, we have

$$^tP(\xi', \tau)\hat{\phi}(\xi', \tau) = \int_{\underline{R}^n} [\exp - i(\langle x', \xi'\rangle + t\tau)](^tP(D)\phi)(x', t)\, dx'\, dt;$$

hence, we get the inequality

$$\sup_{|\tau-\mu|\le 1} |^tP(\xi', \tau)\hat{\phi}(\xi', \tau)| \le \sup_{|\sigma|\le 1} \int_{\underline{R}^n} e^{|t\sigma|} |(^tP(D)\phi)(x', t)|\, dx'\, dt. \tag{7.18}$$

Similarly, we can write

$$\xi_j^{n+1}\, {}^tP(\xi',\tau)\hat{\phi}(\xi',\tau) = \int_{\underline{R}^n} [\exp - i(\langle x',\xi'\rangle + t\tau)](D_j^{n+1}\, {}^tP(D)\phi)(x',t)\, dx'\, dt,$$

$$1 \le j \le n-1.$$

and

$$\tau^{n+1}\, {}^tP(\xi',\tau)\hat{\phi}(\xi',\tau) = \int_{\underline{R}^n} [\exp - i(\langle x',\xi'\rangle + t\tau)](D_t^{n+1}\, {}^tP(D)\phi)(x',t)\, dx'dt$$

and derive the inequalities

$$\sup_{|\tau-\mu|\le 1} |\xi_j^{n+1}\, {}^tP(\xi',\tau)\hat{\phi}(\xi'\tau)| \le \sup_{|\sigma|\le 1} \int_{\underline{R}^n} e^{|t\sigma|} |(D_j^{n+1}\, {}^tP(D)\phi)(x',t)dx'dt, \tag{7.19}$$

$1 \le j \le n-1$, and

$$\sup_{|\tau-\mu|\le 1} |\tau^{n+1}\, {}^tP(\xi',\tau)\hat{\phi}(\xi',\tau)| \le \sup_{|\sigma|\le 1} \int_{\underline{R}^n} e^{|t\sigma|} |(D_t^{n+1}\, {}^tP(D)\phi)(x',t)|\, dx'dt.$$

Combining the inequalities (7.15)-(7.20), we obtain the following estimate

$$A \le \sup_{|\sigma|\le 1} \int_{\underline{R}^n} e^{|t\sigma|} \{|{}^tP\phi| + \sum_{j=1}^{n-1} |D_j^{n+1}\, {}^tP\phi| + |D_t^{n+1}\, {}^tP\phi|\}\, dx'dt$$

hence, by (7.14), we get

$$|\phi(0)| < M \sup_{|\sigma|\le 1} \int_{\underline{R}^n} e^{|t\sigma|} \left\{|{}^tP\phi| + \sum_{j=1}^{n-1} |D_j^{n+1}\, {}^tP\phi| + |D_t^{n+1}\, {}^tP\phi|\right\} dx'dt \tag{7.21}$$

4. Consider, now, on ${}^tPC_c^\infty$ the topology induced by C_c^{n+1}. If ${}^tP\phi_j \to 0$ in C_c^{n+1}, this means that all ${}^tP\phi_j$ have support contained in a fixed compact set of $\underline{R}^n$ and that all the derivatives of order $\le n+1$ of ${}^tP\phi_j$ converge uniformly to zero. Hence, the right-hand side of (7.21) must converge to zero, which implies that $\phi_j(0) \to 0$. But this shows, precisely, that the linear functional

$$\psi \in {}^tPC_c^\infty \to \phi(0) \in \underline{C}, \text{ with } {}^tP\phi = \psi,$$

is continuous on ${}^tPC_c^\infty$ equipped with the topology induced by C_c^{n+1}. Q.E.D.

PROBLEMS

1. Prove that if $n = 2$ $E = (1/2)\pi \ell og\, r$ is a fundamental solution of Δ.

2. Show the equivalence of the ellipticity conditions (E) and (E').

3. Let $\phi \in S(\underline{R}^n)$ and let $u_\phi(r)$ be defined by (7.2). Prove that: (i) $u_\phi(r) \in S(\underline{R}^n)$, $\forall\ \phi \in S(\underline{R}^n)$; (ii) the map $\phi \in S(\underline{R}^n) \to u_\phi(r) \in S(\underline{R})$ is a continuous one.

4. Prove formula (7.7).

5. Compute the Fourier transform of the distribution U defined by (7.8).

6. Prove that if $P = P(D)$ is a partial differential operator with constant coefficients (not all equal to zero), then the only $T \in E'$ such that $PT = 0$ is the distribution identically zero. (*Hint:* Use the Paley-Wiener-Schwartz theorem.)

REFERENCES

[1] Barros-Neto, J., "Kernels associated to general elliptic problems," *J. Funct. Anal.*, 3 (2), 1973-192 (1969).

[2] Bochner, S., and K. Chandrasekharan, "Fourier transforms," *Annals Math. Studies*, No. 19, Princeton University Press, Princeton, N. J., 1949.

[3] Bochner, S., and W. T. Martin, *Functions of Several Complex Variables*, Princeton University Press, Princeton, N. J., 1948.

[4] Bourbaki, N., *Topologie Générale*, Hermann, Paris, 1953-1961.

[5] Bourbaki, N., *Integration*, Hermann, Paris, 1952, 1956, 1959, 1963.

[6] Bourbaki, N., *Espaces Vectoriels Topologiques*, 2 Vols., Hermann, Paris, 1953, 1955.

[7] Buck, R. C., *Advanced Calculus*, 2nd ed., McGraw-Hill, New York, 1965.

[8] Ehrenpreis, L., "Solutions of some problems of division, I," *Am. J. Math.* 76, 883-903 (1954).

[9] Gelfand, I. M., and G. E. Shilov, *Generalized Functions*, Vols. 1, 2, 3, Academic Press, New York.

[10] Gelfand, I. M., and N. Ya. Vilenkin, *Generalized Functions*, Vol. 4, Academic Press, New York.

[11] Gelfand, I. M., M. I. Graev, and N. Ya Vilenkin, *Generalized Functions*, Vol. 5, Academic Press, New York.

[12] Grothendieck, A., *Espaces Vectoriels Topologiques*, 2nd ed., Sociedade de Matemática de S. Paulo, S. Paulo, 1958.

[13] Grothendieck, A., "Sur certains espaces de fonctions holomorphes I," *J. reine angew, math.*, 192 (1), 35-64 (1953).

[14] Hörmander, L., *Linear Partial Differential Operators*, Springer-Verlag, Berlin, 1963.

[15] Hörmander, L., *An Introduction to Complex Analysis is Several*

Complex Variables, Van Nostrand, Princeton, N. J., 1966.

[16] Hörmander, L., "On the division of distributions by polynomials," *Ark. Mat.*, 3; 555-568 (1958).

[17] Horvath, J., *Topological Vector Spaces and Distributions*, Addison-Wesley, Reading, Mass., 1966.

[18] Kelley, John L., *General Topology*, Van Nostrand, Princeton, N. J., 1955.

[19] Kothe, G., *Topologische Lineare Raume*, Springer-Verlag, Berlin, 1960.

[20] Lions, J. L., *Problèmes aux Limites dans les Equations aux Dérivées Partielles*, 2nd ed., Les Presses de L'Université de Montréal, 1965.

[21] Malgrange, B., "Existence et approximation des solutions des équations aux dérivées partielles et des équations de convolution," *Ann. Inst. Fourier, Grenoble*, 6, 271-355 (1955-56).

[22] Nachbin, L., *Lectures on the Theory of Distributions*, Instituto de Fisica e Matemática, Universidade do Recife, Pe., Brazil, 1964.

[23] Peetre, J., "Une caractérisation abstraite des opérateurs différentiels," *Math. Scand.* 7, 211-218 (1959).

[24] Peetre, J., "Rectification à l'article 'Une caractérisation abstraite des opérateurs différentiels', *Math. Scand.* 8, 116-120 (1960).

[25] Royden, H. L., *Real Analysis*, 2nd ed., Macmillan, New York, 1968.

[26] Rudin, W., *Principles of Mathematical Analysis*, 2nd ed., McGraw-Hill, New York, 1964.

[27] Rudin, W., *Real and Complex Analysis*, McGraw-Hill, New York, 1966.

[28] Schwartz, L., *Théorie des Distributions*, I, II, 2nd ed., Hermann, Paris, 1957.

[29] Schwartz, L., *Seminaire* 1954/55, *Équations aux Dérivées Partielles*, Faculté des Sciences de Paris, Sécretariat Mathématique, 1955.

[30] Schwartz, L., *Méthodes Mathématiques pour les Sciences Physiques*, Hermann, Paris, 1961.

[31] Treves, J. F., *Linear Partial Differential Equations with Constant Coefficients*, Gordon and Breach, New York, 1966.

[32] Treves, J. F., *Topological Vector Spaces, Distributions and Kernels*, Academic Press, New York, 1967.

[33] Zygmund, A., *Trigonometrical Series*, I, II, Cambridge University Press, New York, 1959.

AUTHOR INDEX

Numbers in parentheses are reference numbers and indicate that an author's work is referred to although his name is not cited in the text. Underlined numbers show the page on which the complete reference is listed.

SUBJECT INDEX